BABIES, KIDS AND DOGS

Creating a safe and harmonious relationship

Melissa Fallon & Vickie Davenport

Hubble & Hattie

The Hubble & Hattie imprint was launched in 2009 and is named in memory of two very special Westie sisters owned by Veloce's proprietors. Since the first book, many more have been added to the list, all with the same underlying objective: to be of real benefit to the species they cover, at the same time promoting compassion, understanding and respect between all animals (including human ones!) All Hubble & Hattie publications offer ethical, high quality content and presentation, plus great value for money.

More books from Hubble & Hattie –

Among the Wolves: Memoirs of a wolf handler (Shelbourne)
Animal Grief: How animals mourn (Alderton)
Babies, kids and dogs – creating a safe and harmonious relationship (Fallon & Davenport)
Because this is our home ... the story of a cat's progress (Bowes)
Boating with Buster – the life and times of a barge Beagle (Alderton)
Bonds – Capturing the special relationship that dogs share with their people (Cukuraite & Pais)
Camper vans, ex-pats & Spanish Hounds: from road trip to rescue – the strays of Spain (Coates & Morris)
Cat Speak: recognising & understanding behaviour (Rauth-Widmann)
Charlie – The dog who came in from the wild (Tenzin-Dolma)
Clever dog! Life lessons from the world's most successful animal (O'Meara)
Complete Dog Massage Manual, The – Gentle Dog Care (Robertson)
Detector Dog – A Talking Dogs Scentwork Manual (Mackinnon)
Dieting with my dog: one busy life, two full figures ... and unconditional love (Frezon)
Dinner with Rover: delicious, nutritious meals for you and your dog to share (Paton-Ayre)
Dog Cookies: healthy, allergen-free treat recipes for your dog (Schöps)
Dog-friendly Gardening: creating a safe haven for you and your dog (Bush)
Dog Games – stimulating play to entertain your dog and you (Blenski)
Dog Relax – relaxed dogs, relaxed owners (Pilguj)
Dog Speak: recognising & understanding behaviour (Blenski)
Dogs on Wheels: travelling with your canine companion (Mort)
Emergency First Aid for dogs: at home and away Revised Edition (Bucksch)
Exercising your puppy: a gentle & natural approach – Gentle Dog Care (Robertson & Pope)
For the love of Scout: promises to a small dog (Ison)
Fun and Games for Cats (Seidl)
Gods, ghosts, and black dogs – the fascinating folklore and mythology of dogs (Coren)
Helping minds meet – skills for a better life with your dog (Zulch & Mills)
Home alone – and happy! Essential life skills for preventing separation anxiety in dogs and puppies (Mallatratt)

Know Your Dog – The guide to a beautiful relationship (Birmelin)
Life skills for puppies – laying the foundation for a loving, lasting relationship (Zuch & Mills)
Living with an Older Dog – Gentle Dog Care (Alderton & Hall)
Miaow! Cats really are nicer than people! (Moore)
Mike&Scrabble – A guide to training your new human (Dicks & Scrabble)
My cat has arthritis – but lives life to the full! (Carrick)
My dog has arthritis – but lives life to the full! (Carrick)
My dog has cruciate ligament injury – but lives life to the full! (Haüsler & Friedrich)
My dog has epilepsy – but lives life to the full! (Carrick)
My dog has hip dysplasia – but lives life to the full! (Haüsler & Friedrich)
My dog is blind – but lives life to the full! (Horsky)
My dog is deaf – but lives life to the full! (Willms)
My Dog, my Friend: heart-warming tales of canine companionship from celebrities and other extraordinary people (Gordon)
No walks? No worries! Maintaining wellbeing in dogs on restricted exercise (Ryan & Zulch)
Partners – Everyday working dogs being heroes every day (Walton)
Smellorama – nose games for dogs (Theby)
Swim to recovery: canine hydrotherapy healing – Gentle Dog Care (Wong)
A tale of two horses – a passion for free will teaching (Gregory)
Tara – the terrier who sailed around the world (Forrester)
The Truth about Wolves and Dogs: dispelling the myths of dog training (Shelbourne)
Unleashing the healing power of animals – True stories abut therapy animals – and what they do for us (Preece-Kelly)
Waggy Tails & Wheelchairs (Epp)
Walking the dog: motorway walks for drivers & dogs revised edition (Rees)
When man meets dog – what a difference a dog makes (Blazina)
Winston ... the dog who changed my life (Klute)
The quite very actual adventures of Worzel Wooface (Pickles)
Worzel Wooface: The quite very actual Terribibble Twos
You and Your Border Terrier – The Essential Guide (Alderton)
You and Your Cockapoo – The Essential Guide (Alderton)
Your dog and you – understanding the canine psyche (Garratt)

Disclaimer

Please note that no dog was deliberately frightened or worried during photographic sessions. Any images used to depict dogs feeling uncomfortable about specific situations were taken opportunistically, whilst the dogs were engaging with their surroundings.

While the authors and publisher have designed this book to contain up-to-date information regarding the subject matter, readers should be aware that information is constantly evolving. Having put into practice the advice in this book, if you are at all worried about your dogs behaviour, it is strongly recommended you seek the advice of a professional behaviourist on veterinary referral. The authors and publisher shall have neither liability nor responsibility with respect to any loss, damamge or injury caused, or alleged to be caused, directly or indirectly by the information contained in this book.

www.hubbleandhattie.com

For post publication news, updates and amendments relating to this book please visit www.hubbleandhattie.com/extras/HH4890

First published September 2016 by Veloce Publishing Limited, Veloce House, Parkway Farm Business Park, Middle Farm Way, Poundbury, Dorchester, Dorset, DT1 3AR, England. Fax 01305 250479/email info@hubbleandhattie.com/web www.hubbleandhattie.com ISBN: 978-1-845848-90-3 UPC: 6-36847-04890-7 © Melissa Fallon, Vickie Davenport & Veloce Publishing Ltd 2016. All rights reserved. With the exception of quoting brief passages for the purpose of review, no part of this publication may be recorded, reproduced or transmitted by any means, including photocopying, without the written permission of Veloce Publishing Ltd. Throughout this book logos, model names and designations, etc, have been used for the purposes of identification, illustration and decoration. Such names are the property of the trademark holder as this is not an official publication.
Readers with ideas for books about animals, or animal-related topics, are invited to write to the editorial director of Veloce Publishing at the above address. British Library Cataloguing in Publication Data – A catalogue record for this book is available from the British Library. Typesetting, design and page make-up all by Veloce Publishing Ltd on Apple Mac. Printed in India by Replika Press.

Contents

Acknowledgements • Dedication • Introduction 4
Foreword by Steve Mann 6

1: Assessing your dog 7
    Emotional behaviour 7
    What causes a dog to bite 8
    Potential problem behaviours/scenarios . . .8
    Observational skills 15
    Pre-baby/child assessment 17

2: Preparation. 22
    Preparing your dog 22
    Build confidence 22
    Walks/exercise. 24
    Flexible routine/energy. 26
    Avoiding temper tantrums 26
    To chew or not to chew? 27
    Establish boundaries 28
    Exercises 1-12 30

3: Introducing the new family member 50
    Protocol for the first few meetings 50
    Going forward.. 51
    Important points 55

4: Toddler training 56
    Happy dog vs unhappy dog 56
    Toddler touches 56

Respecting your dog's food/resources. 60
Let sleeping dogs lie … 60

5: Raising children around dogs 73
    Unfamiliar dogs 73
    Familiar dogs 73
    Bringing a new dog into the home 74
    Educate 75
    Settle in 76
    Get your children involved 76
    Basic training 76
    Trick training. 80
    Exercise. 78
    Games 79
    Messy play.. 85
    Kong™ recipe ideas 85

Appendix: Resources and further reading 91
Index 92

Acknowledgements • Dedication • Introduction

Acknowledgements

We would like to thank all of the people and dogs in our lives who have helped us to develop our knowledge and skills within this field. We would also like to thank Annemarie, 'the phodographer,' for helping us to capture some of our images, and Natasha Thompson for developing our illustrations of 'Charlie and Champ.' Finally, special thanks go to our volunteer proof readers for providing the time to constructively critique our work.

Dedication

We dedicate this book to all loyal dogs and gorgeous children, who provide us with both challenges and joyous memories.

Introduction

There are many benefits for both children and dogs who are raised together in a safe and happy relationship. Dogs can develop their confidence and become better socialised, and will therefore be more comfortable in social and novel situations. This also means that they are less likely to react negatively to either familiar or unfamiliar people. The numerous benefits for children include more positive relationships, a reduction in potential health problems, increased exercise, and confidence-building. Being less fearful of dogs, in general (and other animals, too, hopefully), is obviously extremely beneficial in society.

Studies reveal that owning pets increases immunity, and being around dogs in particular can teach children about the importance of exercise, and help promote a healthy and active lifestyle. And, of course, the benefits to mental health are also well known: interaction with animals reduces stress for both dog and child, and can promote

Building confidence in dogs will encourage him to relax in new situations and around new people.

Having dogs as part of the family fosters healthy lifestyles.

calm, relaxed, happy households. Dogs can provide security and companionship, as well as help teach responsibility and empathy.

Only ever offer a home to a dog after long and careful thought and consideration of all that will be involved in doing so: owning a dog requires care and responsibility, as well as a financial commitment. In an ideal world, all dogs would be child-friendly, have undertaken a minimum amount of training, and have the necessary background and upbringing to allow them to develop appropriate manners to deal with being around a variety of people and situations. This is even more important if children will become part of the family at a later date. If you do not feel that your dog is ready for this huge change in circumstance, this book can help you to work on positively preparing him and prevent potential problems.

Practical guidance on how to prepare your dog for being around children – and how to supervise him when they are together – is coupled with a knowledge and understanding of dog behaviour, and the ability to pick up on subtle cues that all is not well. In order to properly supervise just such a situation, it is essential to have certain basic skills. Although parents should aim to be present at all times when children and dogs are together, this is not always possible, and so guidance is provided about how to create safe and effective barriers between your dog and your children on occasions such as this.

A certain level of confidence that both child and dog are calm and at ease in each other's presence, and know how to interact appropriately with each other, is necessary, and should as a result of educating your child and appropriate training for your dog, which should allow him to interact safely, as well as behave appropriately in situations in which he is uncomfortable. He should also have the space and freedom to be able to move away from situations that cause him concern, so that he feels in control of his environment.

As already noted, dog ownership should not be taken lightly, and neither, of course, should your child's safety. Even

Children and dogs can provide each other with companionship and true friendship.

if you do not actually have a dog in your home, but your child may come into contact with one who belongs to a friend, say, or someone else in your family, it is important to prepare him to meet your child, or your new arrival.

Please note, this book is not intended as an aid for any serious behavioural problems your dog may have, which should be referred to a suitably qualified professional.

For ease of reading, the text refers to the dog as a male throughout: however, female is implied at all times.

Foreword

Contrary to the popular showbiz saying, 'never work with children and animals,' I'd go with quite the opposite: never *stop* working with children and animals: what could possibly be more important!?

The authors, Melissa and Vickie, have put together this brilliant, beautifully-illustrated book, which serves not only as a reference for owners and parents, but also as a training 'go-to' to ensure that the 'family' has the happiest, safest and empathetic relationship with their dog possible.

Dogs and people have lived together harmoniously for thousands of years, but never before have the day-to-day social, work and time pressures been more taxing for owners of dogs. Throw a child or two into that mix and we can see why we need this book!

Safety, enjoyment and quality of life should never be compromised, for parent, for child, or for dog.

First impressions last, and *Babies, kids and dogs* endeavours to make sure you 'do it once, and do it right.'

Introductions, assessments, confidence-building and training are all covered in such a clear way that all the family can have access to really understanding just what dogs need and want, in order to enjoy their life to the full as part of the family.

'Part of the family,' that's a term often used when people refer to their dog, but how often is it actually true?

On many occasions 'Mum' (like it or not!) becomes the main Carer, Trainer and Picker-upper of the 'family' dog. Within the pages of this book are several chapters that actively and positively encourage everyone to get on board with the family dog, 'their' family dog, covering topics such as how to manage interactions between toddlers and dogs, key questions and guidance for parents to discuss with their children, and my personal favourite, activities that older children can get involved in with their dog.

If we can get children interacting positively with dogs in a loving, safe and empathetic way, it can only bode well for the future.

It's our responsibility.

Enjoy the book, your dog, and your journey.

Steve Mann
Founder of The Institute of Modern Dog Trainers
www.imdt.uk.com

1: Assessing your dog

Assessing your dog for potentially aggressive behaviour is the first and most important consideration when integrating him with children. A qualified behaviourist can carry out a pre-baby assessment for you, but it's important to develop your own observational skills and understanding of what motivates a dog to behave in a certain way.

Understanding why your dog behaves the way he does will help you to understand his needs, and assist in building positive relationships within the family.

Prevention is better than cure, of course, and prior to introducing your dog to babies and children, it's essential that you know and understand his temperament, as this will help to better predict his behaviour, deal with risk factors, and avoid problems arising in the first place, and he will come to understand that babies and children are not a threat to him, or something to be fearful of. Secondly, important though it is to always supervise him around children, it is even more important to be able to read his behaviour correctly, no matter how subtle the signals might

be, as understanding his behaviour and knowing how to respond appropriately can prevent potential issues.

Emotional behaviour
Behaviour can have an emotional element to it. Consider examples of highly emotional responses in humans: the death of someone close to us, for example, will cause us to be upset; losing or being deprived of something valuable could cause frustration; a break-in or attack will provoke a fear response, and an argument or disagreement may cause us to become angry. Certain situations can give rise to a more emotional response, which will, in turn, affect our behaviour. Emotional responses will exaggerate behaviour, making it more intense, unpredictable, and more varied (a range of behaviours may be demonstrated over a short period of time).

The adaptive value of these emotional behaviours is what helps us to survive, and occurs in situations such as –

- protecting us and our dependants from being hurt
- protecting resources (money or food)
- escaping from a dangerous situation
- avoiding situations that we have learnt can cause pain or affect health

Animals are no different to us in this respect, in that they also use adaptive responses in order to ensure survival, and dogs exhibit these innate responses to defend resources, and protect themselves and their young from harm, whilst learning to avoid harmful situations in the process.

Many factors can cause dogs to become aggressive, including temperament, which is not necessarily determined by breed. Many studies have shown that animals can develop a cognitive bias – a systematic error in thinking that affects decisions and judgements – which is sometimes related to how events are remembered.

A lack of communication – not being understood – can also cause frustration on both sides. For example, your dog will not automatically know what you want from him when you ask him to 'sit,' and even though constantly repeating the word may sometimes appear to be successful, in a lot of cases this is simply because the dog has become tired enough to *want* to sit down, and the word and resultant behaviour has become associated by chance only.

What causes a dog to bite

The best way to prevent your dog feeling the need to resort to biting is to understand how he is feeling around children and other stimuli, which should then allow you to –

- adapt or manage the environment so that he feels able to relax
- respond with alternative behaviours
- change his emotional state and perception of the situation through training

Chapter 2 provides advice on how to prepare your dog to be around children, and avoid some of the situations – as outlined below – where he may be more likely to bite. Usually, situations like this are driven by underlying emotions, which will vary in each dog. Bear in mind, too, that practicing a particular behaviour – including aggressive behaviour – can mean that, over time, this becomes your dog's default behaviour in particular situations.

Potential problem behaviours/scenarios

- predatory behaviour
- seeking attention
- frustration
- conflict anxiety
- pain (or fear of pain/previous negative experience)
- fear of the unknown
- fear of losing a reward
- play

PREDATORY BEHAVIOUR

Usually motivated by appetite, the instinct to chase is reinforcing in itself, so is not always caused by the need to feed. Predatory behaviour includes approaching the perceived prey item, followed by behaviours such as stalking, crouching, or staring, with ears flat and tail low; very still or moving slowly and low to the ground. These behaviours will usually be seen directed towards other animals, and, although usually instinctive, can be prompted by various things, and can be difficult to deal with. Be able to recognise predatory behaviour, though do not assume that just because your dog has a high chase/prey drive he will exhibit predatory behaviour within the household, and particularly toward children.

SEEKING ATTENTION

Some dogs go to great lengths in order to gain attention. Attention-seeking behaviour might start with your dog nudging you, or barking, which, if ignored, may prompt him to try alternative techniques, such as nipping or pawing. This behaviour can pose a risk around young children as it becomes progressively more pronounced, and pawing your

---

INTRODUCING A CUE

When training, try to ensure that you –
Avoid using the same cue word to prompt multiple different behaviours: for example, using the 'down' cue to ask your dog to lie down on the floor as well as get off the furniture, or to prevent him jumping up.

Be consistent with your use of verbal and visual cues (hand signals). The verbal cue should only be introduced once your dog is regularly offering the behaviour that you want. In this way, he will associate the behaviour that he has already learnt with the verbal cue. This will also need to be repeated with reinforcement (whatever is motivating for him), until the new behaviour becomes a habit. The reinforcement could be a specific toy, praise or type of food. If using food, you may need to start off using high value rewards, such as small amounts of cheese, ham or sausage, gradually introducing a variety of high and low value treats (dry biscuits, say).

Be careful of dogs who offer their paw as a way of gaining attention, as this could inadvertently injure young children who are not prepared for this.

legs for attention could also be problematic if, say, you are holding a baby.

Attention-seeking behaviour can vary greatly, and will usually be tried on everyone; not just children. It's usually the case that the behaviour develops because your dog is seeking some kind of reward, whether it be food, play or attention, and, if successful, then becomes the learnt behaviour. This may also change over time, going from nudging, whining, nibbling and barking to snapping, biting, and growling, dependent on if, how, and when the behaviours are reinforced.

Successfully dealing with attention-seeking behaviour includes teaching a more acceptable alternative behaviour – asking politely for attention, such as a sit to greet, for example. Simply ignoring the undesirable attention-seeking behaviour should be done carefully as your dog may become frustrated at being ignored, which is why the appropriate behaviour should be trained alongside. Remember, too, that even reacting negatively may reinforce the behaviour, as your dog is still receiving the attention he seeks.

Frustration

Aggressive behaviour can be caused by frustration, which can develop in various situations, though most usually when your dog is trying to obtain something. In this situation he may redirect his frustration towards whoever is close by – children, adults or animals. Children can also be the cause of his frustration by walking near him with food; taking away his toys; suddenly touching or grabbing him, and not giving him the space he needs to move about and sleep or rest in peace. Building tolerance to frustration can be achieved through self-control, as discussed in Chapter 2.

Conflict anxiety

This can occur when your dog may be in conflict about something within his environment, and unsure of whether to approach or avoid it; you may see uncertainty in his movements as he approaches and cautiously sniffs before hesitantly backing away. He may actually want to interact with whatever it is, but is unsure and nervous about doing so.

Conflict behaviour can incorporate quite unpredictable behaviour. He should be allowed to build confidence in his own time: use gentle positive training to help him to do this, rewarding him when he summons up sufficient confidence.

Pain (or fear of pain/past negative experience)

If experiencing pain in any part of his body, a dog may respond with stress signals (lip-licking, yawning, moving away), or warning signals (baring his teeth or growling), if touched there. Unexpected pain can develop at any stage in a dog's life, especially as he ages, and should be attended to immediately.

If he is generally quite comfortable with being touched, and happy in his surroundings, he is more likely to respond by trying to move away from the contact, or with a warning signal, like a growl, as a first line of defence that indicates he does not like being touched, and would like some space. If his warning signals (identified as the lower levels of aggressive behaviour) have previously elicited the response he seeks, he will rely on them doing so again, with more aggressive behaviour becoming unnecessary or a very last resort. Of course, how he reacts will depend to a certain extent on the level of pain he is experiencing.

Be aware that even though his pain may be localised, he may still not appreciate being touched elsewhere on his body. Previous experiences of being handled whilst in pain may have made him sensitive to being handled generally, as

Babies, kids and dogs

Managing a dog who's reactive whilst on the lead

If your dog barks and lunges at particular stimuli – other dogs, animals, bikes, cars, people, etc – this will compromise loose lead training, and also pose a risk when walking with children. If he does exhibit reactive behaviour, it's best to not walk him alongside your toddler or a pram, as the behaviour can be redirected toward them due to frustration at not being able to get to the reason for his reactivity. This behaviour is best dealt with by a professional, but the following advice may also prove helpful.

Using distraction, or teaching your dog to 'focus on me' may help manage situations when he is reactive. While out and about, identify in advance stimuli that he may react to, and distract him from this by feeding him treats as you walk past the stimuli.

Teaching him a good 'focus on me' cue can be a useful aid in many different situations, and especially with a reactive dog (see also accompanying photos on opposite page). As with all training, keep the sessions short in order to maximise learning potential, and make the activity interesting for your dog. If he appears to lose interest or shows signs of becoming frustrated, give him a treat and stop the session, resuming on another day –

- Work in an area where there are few distractions – in one room, or the garden, for example.
- Get his attention and ask him to sit, if he knows this cue (standing or laying are fine as well). Show him a treat in your hand, then slowly raise your hand to your eye level (if there's a chance that your dog may jump up, hold the treat away from your face). As soon as he looks into your eyes (or at your face if he does not want to make eye contact), say 'good' and reward him with the treat. Repeat this until he is reliably (8 out of 10 times) looking at your face each time you lift your hand to eye level.
- Once he is at this point, you can begin to introduce your verbal cue (choose any word you like and use this consistently). Repeat the steps above, but give your cue just before you raise your hand, establish he is looking where you want him to, praise him and reward him.
- Repeat this to the point where you can omit the treat from your hand, simply give the verbal cue, lift your hand to eye level, wait for the required response from your dog, say 'good' and reward him.
- Continue to practice this in the area of low distraction, and gradually phase out the hand cue in stages – not taking the hand all the way up to your face, for example – until your dog responds to just your cue.
- Gradually introduce distractions, such as people walking past you, or a ball being thrown in the vicinity, though note it will probably be necessary to go back over a few steps each time you do.
- Don't give the cue too many times without the correct response. After asking twice, go back to *showing* him what you want him to do. If you repeat the cue too many times with no response, he will simply learn to ignore it.

Once you have a 'focus on me' cue that your dog reliably responds to, when on walks, use the cue to distract him from anything he may react to. If you spot, say, another dog in the distance, use the 'focus on me' cue and reward your dog, then allow him to look at the other dog. Once he has done so, give the cue again and reward him for complying. If you repeat this each time you are out on a walk, your dog should soon learn that it is in his interest to look at you rather than react to the stimuli. As his owner, it is your responsibility to properly assess each situation, and decide what he can cope with as well as react accordingly: you might decide to move him away from the stimuli, or use more encouragement to walk him past it.

Ask him to sit and then show him the treat in your hand.

Raise your hand to your eye level and hold the hand quite close to your eye to encourage him to return your gaze, or at least look at your face.

Once he understands where he is supposed to look, move your hand away from your face.

he has established an association between feeling pain and being touched – even if he is no longer *in* pain.

So even though he may have been given the all-clear by his vet, a previous experience of pain in that location may cause him to be extra sensitive. In order to overcome his fear of pain and being handled, it will be necessary to desensitise him to touch – all over his body, but especially in that area of his body recently affected. The handling protocol as described in Chapter 2 can help with this.

As your dog ages, it may be necessary to make changes in the way children interact with him. He may become less tolerant of noise, for example, and be more reactive to it. He also may be suffering from canine cognitive decline (CCD), which can cause him to become disorientated and confused. His becoming more prone to aches and pains is also likely to occur, so give him the space and calm he needs.

FEAR OF THE UNKNOWN
Lack of socialisation during their young years is a common cause of fear in dogs, as they simply do not know how to

The Canine 'Ladder of Aggression

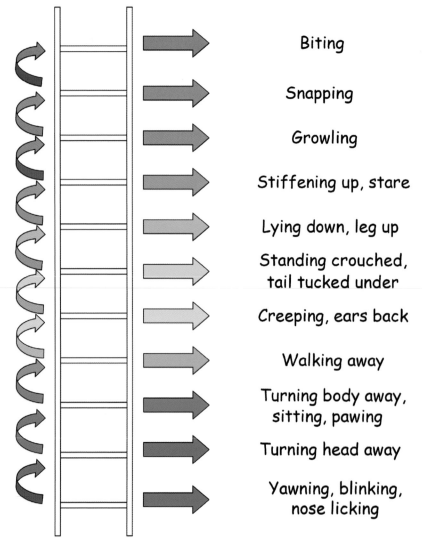

Biting

Snapping

Growling

Stiffening up, stare

Lying down, leg up

Standing crouched, tail tucked under

Creeping, ears back

Walking away

Turning body away, sitting, pawing

Turning head away

Yawning, blinking, nose licking

How a dog reacts to stress or threat

© Kendal Shepherd 2004

THE CANINE LADDER OF AGGRESSION

The Ladder of Aggression is a depiction of the behaviours a dog may exhibit in response to an escalation of perceived stress/threat, from very mild social interaction and pressure (to which blinking and nose-licking are appropriate responses), to severe situations, when overt aggression may be shown. The purpose of such behaviour is to deflect a threat and restore harmony, and the presence of appeasing and threat-averting behaviour in the domestic dog's repertoire is essential to avoid the need for potentially damaging aggression. The dog is a social animal for whom successful appeasement behaviour is highly adaptive, and this is used continually and routinely in everyday life.

It is most important to realise that these gestures are simply a context- and response-dependent sequence which will culminate in threatened or overt aggression only if all else fails. Contrary to persistent misinformation, the gestures identified are nothing to do with a purported dominant or submissive state relative to companions. In all dogs, inappropriate social responses to appeasement behaviour will result in its devaluing, and the necessity, from a dog's perspective, to move up the ladder. Aggression is therefore created in any situation where appeasement behaviour is chronically misunderstood, and not effective in obtaining the socially expected outcome. Dogs may progress to overt aggression within seconds during a single episode if the perceived threat occurs quickly and at close quarters, or learn to dispense with lower rungs on the ladder over time, if repeated efforts to appease are misunderstood and responded to inappropriately. As a consequence, a so-called 'unpredictable' aggressive response, without any obvious preamble, may occur in any context which predicts inescapable threat to the dog, when, in reality, it was entirely predictable.

(K Shepherd, 2009. BSAVA Manual of Canine and Feline Behaviour, 2nd edition, pages 13-16. Editors Debra F Horwitz and Daniel S Mills)

The diagram, opposite, demonstrates typical behaviours a dog may exhibit when trying to avoid or remove himself from a situation he finds stressful or threatening. If low-level calming signals such as yawning or looking away do not produce the desired result, he may feel his has no option other than to escalate his behaviour, as per the diagram. If he has learnt from past experience that his low-level signals are not acted on, he may not bother with these at all, and may go straight to higher level behaviours such as growling, snapping, lunging, and even biting. Quite apart from this, it's important to understand that not all dogs will show all of the stages featured before biting, as their response depends on environment, previous experience and context, and individual temperament.

Building confidence in dogs encourages them to relax and settle in new situations and around new people and objects.

behave in certain situations. It is a natural adaptation to be fearful or cautious of the unknown, as this can be dangerous and affect health, and even survival. Your dog needs to learn that novelty is not something to be nervous of, and, if anything, he should regard it as a good thing. Children, and people in general, should have positive associations for your dog, and not be regarded as threatening or a source of fear.

Canine fear responses and stress signals include avoidance, lip-licking, turning away, bulging eyes, tense face and body (mainly noticeable around the mouth), crouching, cringing, and low ears and tail (dependent on breed type).

If he has not been socialised around children; has learnt that he cannot escape the attentions of children; has no confidence around children and their sometimes unpredictable behaviour, or is expected to accept situations that he finds frightening, and is punished for giving warning signals, your dog is more likely to resort to heightened aggressive behaviour in situations he finds difficult to cope with. It is important for him to always have a safe haven (his bed, ideally), and not have to resort to finding a bolthole, such as under a table or bed, which then assume negative associations.

Formal training can help your dog to develop positive associations with children, and talking with your children will educate them about the right way to behave around dogs (discussed in Chapter 4).

Of course, being nervous of touch in certain areas of his body may not always be due to pain. This could be due to a lack of positive handling when a puppy, or simply that some areas of his body are more sensitive to touch (usually paws or genital area). Desensitisation and positive handling can again help here; building this very slowly, dependent on your dog's prior exposure and severity of fear.

FEAR OF LOSING A REWARD

Also referred to as 'resource-guarding' (see page 14) this is when a dog is feeling fearful of losing a resource that he perceives as his: food, a toy, his bed. In this instance, he may try and increase the distance between whoever or whatever he considers is threatening to take his resource. If none of his distance-increasing behaviours work (teeth-baring (agonistic pucker), low growl, barking, crouching), he may feel he has no other option than to escalate the behaviour.

PLAY

We expect an awful lot from our dogs when it comes to bite inhibition (sometimes referred to as having a soft mouth, whereby the dog learns to moderate the strength of his bite), which is an important factor in their socialisation. Dogs do seem to manage this quite well, however, and even, in some cases, apparently discriminate between dogs and people, although further discrimination between people and children is a lot to ask.

Rough-and-tumble play between dogs can be very difficult to read and interpret to the untrained eye, and play between people and dogs even more so. When two dogs play together (and particularly if they do not know each other), owners sometimes worry that the play is too aggressive, especially when a much larger, boisterous

Babies, kids and dogs

Dealing with resource-guarding

Resource-guarding can be a common problem amongst dogs. The simple training pointers given below can help teach your dog that giving up a resource leads to receiving something better, so he will not feel the need to try and hold on to what he has. It's important for all dogs to learn to give up objects when asked, and exercise 8 in Chapter 2 can help with this.

Your dog should be able to enjoy his food in peace, and so, wherever possible, keep children away from him while he is eating. However, teach him to expect good things if you touch his food bowl or other resources to avoid him developing resource-guarding behaviour, or if you are concerned about being able to manage feeding times safely. Your objective should be to get to the stage where your dog is comfortable giving up high-value resources, such as treats, toys and food bowls. To help with this, try feeding small amounts of food often, so that as he finishes each 'meal', you take the bowl and add a few more treats to it before placing it back on the floor for him to enjoy the contents.

If he is already showing some possessive behaviours, ensure he is left alone when he has resources that he wants to guard. If it is food, for example, use a baby gate to separate him when he is eating, or feed him in the garden, weather-permitting. This is only a short-term safety measure, however, and the problem should be properly addressed as soon as possible. Certainly, if he is showing severe resource-guarding behaviour, such as growling, snarling, snapping, etc, it's strongly recommended that you seek the help of a qualified behaviourist without delay.

- Punishment should not be used on any account, if your dog is already resource-guarding, and displaying the behaviours mentioned. Not only is this a cruel and unnecessary method of control, it can also possibly exacerbate the situation, causing your dog to become even more defensive.
- Avoid supplying high-value resources around children.
- Help your dog to learn that he does not need to fear his resource being taken from him, by exchanging for another resource; perhaps of higher value.
- Resource-guarding can be a difficult problem, which should always be managed and dealt with carefully and slowly, and with the help of a qualified behaviourist.

dog is playing or attempting to play with a much smaller and more nervous animal. Worries such as these – whilst understandable – can cause us to become over-cautious, resulting in frustration for our dog if we prevent him from having fun in this way.

However, due to differences in size and individual temperaments, it's not a good idea to allow him to take part in inappropriate play that may result in problems. Therefore, it is worthwhile developing a good understanding of the play signals that he displays within a play session, such as play bows, reciprocal chase, etc. Having a good understanding of these behaviours can help you to recognise if he uses some inappropriate play techniques, such as body slamming, lunging and snapping, to induce play, which is particularly important to know if he engages

in play with family members and other people. Play is a high arousal activity, which can cause frustration in some dogs, especially if the play is not reciprocal and therefore inappropriate.

Some people encourage inappropriate play behaviour in their dog by playing with them in a rough way, which also encourages 'mouthing' (something a dog will do when he is extremely excited or simply desperate for attention, and in itself not a problem although some dogs bite out of fear or frustration: a type of biting that can indicate problems with aggression). Neither is rough play a problem per se if carefully managed, although dogs are generally good at discriminating between those who enjoy rough play and those who are more likely to prefer a game of fetch. It is, therefore, important to remember that your dog may accidentally nip during play if he is used to practicing this behaviour with certain people. This can also be the case when playing tug with him, when he could try biting the end of the toy that you are holding onto, and accidentally nip.

Be aware of your dog's arousal levels when playing around children, and opt for games that are as calm as possible. Learning appropriate play behaviours as early as possible is important for your dog, as any children and babies he comes into contact with are likely to want to interact with him. Children need to be taught a safe way of interacting with dogs as well, of course, which will not excite them and preclude mouthing.

Tugging can be an appropriate game for an adult or older child to play with your dog, if he has been taught how to play appropriately, and his arousal levels are managed. Ideally, use a long, fleecy tug toy to prevent hands being grabbed.

If your dog becomes very excited or highly aroused very quickly, frustration can result, which could cause him to mouth or lunge at objects. If this behaviour becomes a learnt response every time he becomes excited, it will be necessary to teach him how to control his frustration.

Observational skills

Being able to read and correctly interpret your dog's behaviour is very important, especially when anticipating what he might do next. Try and put yourself in his place and think from *his* point of view. Our dogs don't necessarily *want* to interact with children, or lots of different people,

---

DEALING WITH MOUTHING

Problem-mouthing is always easier to deal with in puppies, who have not had a chance to practice the behaviour as much as older dogs have, and who are, in any case, less likely to have a powerful bite. The following exercises can be used with dogs of any age, however.

Mouthing during play

Play with a long, fleecy tug toy, so that your dog's mouth is not near your hand. As soon as he begins to mouth your arms or hands during play, calmly take the toy from him and walk away. In time, he should realise that this behaviour means an end to playtime. However, be aware that doing this can cause frustration in your dog, and if he is very aroused he may follow, trying to gain your attention by mouthing your legs or biting at your clothing, etc.

Should he react in this way, distract him by making a sudden noise, such as dropping a set of keys onto a hard surface. This is not meant to scare him, but, rather, interrupt his concentration. The noise needs to be just loud enough and novel enough to cause him to stop what he is doing.

When you have his attention again, immediately redirect it to something else that is calmer and quieter – perhaps going out into the garden to toilet.

Mouthing for attention

If your dog engages in random mouthing, and not just during a play session, again, walk away so that he learns that this behaviour is not acceptable. When your dog does seek attention, and wants to interact, teach him an alternative behaviour – to bring you a toy or sit at your feet, say – which elicits the response he seeks, and he has no need to mouth. Every time your dog brings you a toy, always engage with him, or if he comes to you and sits, make a fuss of him to encourage this behaviour.

Should he not react to the noise in the way required, try teaching him a 'leave it' cue, as described in exercise 9 in Chapter 2, as this will also help him become generally less impulsive, as well as teach appropriate play manners. Try and prevent him becoming over-excited during play, and end play with some gentle stroking, or provide him with a chew or a filled Kong™ to settle with (exercise 10, Chapter 2).

Babies, kids and dogs

and it's up to us to notice if he becomes uncomfortable with a situation by observing the signals he is displaying, which typically include –

- yawning
- nose/lip-licking
- turning/looking away
- sniffing the ground
- crouching

His body posture may appear rigid, and he may also be tense around the corners of his mouth. He may hold his tail down and still, though he may also wag it very slowly. He may also begin to pant in an effort to calm himself, although panting can be triggered by multiple causes. A variety of additional resources are available online, as mentioned in Chapter 4, as well as books that can help you recognise these signals.

If your dog is displaying behaviour like this, do not allow children to approach him or invade his space, and maybe even take him out of the situation that he is finding difficult to deal with. If he is showing more obvious signs of

The whites of your dog's eyes may show (whale eye) when he is looking away from an object or situation that is making him ill at ease.

Signs he's unsure include withdrawing from an object or person: a paw lift may also be seen.

Signs of discomfort include ears pulled back, head lowered, and mouth that is tight at the corners. A nose lick may also be seen.

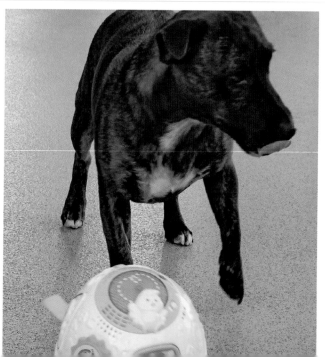

A relaxed body posture can be seen in floppy legs, low, relaxed tail and ears, and no tension lines around the mouth.

Relaxed facial expressions include an open mouth which is relaxed at the corners, and not pulled back over the teeth.

agitation – growling, teeth-baring, pacing – take immediate action to prevent the problem escalating by removing him from the scene. Until your dog has been properly assessed, be cautious of allowing children to interact with him.

A dog who is comfortable and relaxed will appear loose-limbed and fluid in movement. He may hold his tail level to his body, and his facial expressions and the corners of his lips will be relaxed. Holding his tail higher than his body could be an indication that he is interested or excited, so take care that this does not escalate.

Pre-baby/child assessment

It is good to recognise situations where your dog may need more guidance from you in order to cope, which should also help you build on your observational skills, and note subtle behaviours to watch for. It can be very helpful to complete a pre-baby/child assessment that should determine whether or not he feels comfortable in a given situation, and with various stimuli, and will also allow you to note what behaviours he displays when feeling uncomfortable.

You can complete an assessment yourself, using Exercise A at the end of this chapter, or ask a qualified behaviourist to do this for you during a home visit.

WHAT TO ASSESS AND WHY

In order to train effectively, and manage situations in a safe way, it is important to have a good understanding of your dog's temperament, including any learnt behaviours; how he generally responds to given situations, a range of other animals, and a variety of objects, and how quickly he reacts and at what level of arousal.

HOW TO CARRY OUT THE ASSESSMENT

The stimuli/tasks outlined in Exercise A should be set up at

Every dog should be able to relax undisturbed in his bed.

levels that your dog can comfortably deal with, and slowly increased in intensity, ensuring that he remains calm and relaxed. If, at any point, he begins to look anxious or unsure, stop the assessment immediately to prevent him associating the stimuli/task with something worrisome or unpleasant. It's simply not possible to have a dog who is 100 per cent bombproof, but it *is* possible to provide him with sufficient guidance to deal with multiple situations calmly and comfortably.

Consider each of the six essentials in the exercise, and, using the red, amber, green light system, gauge your dog's reaction to each stimuli/task. Remember that every dog is an individual, with his own personality. Some are more socially-inclined and enjoy interaction, whereas others are more reserved. A naturally shy and/or nervous animal will possibly feel more intimidated by children, and this will be evident with more red and amber categories applying to him.

As soon as you detect any sign of discomfort, stop the exercise. Carry out the assessment over a period of time so as not to overwhelm your dog and prevent stressful situations. It should always be your dog's choice whether or not to interact, and he should always be able to move away from a situation.

It's important to keep in mind that, as dogs age, their tolerance levels may also change, especially if they are experiencing pain due to age-related conditions such as arthritis.

If you have multiple dogs or animals within the home, be aware that your dog may behave differently when the other dog(s) or animals are present. It is important to test as many of these exercises as possible in order to get a full understanding of what you need to prioritise when training your dog in preparation of his being around children.

ASSESSMENT CHART
GRADES

The six essentials		Your dog seems to cope well and may require only basic preparation/training	Your dog may experience some difficulties, and may need more work in this area	You may need more time to train your dog: begin at lower levels or at a distance from the stimuli. Take your time and be patient
	STIMULI/TASKS	Assess your dog in the following situations, especially if he has not had exposure to them or you are unsure about how he may react. Always be calm and careful, and allow him the choice to move away. Based on your assessment here, you will be able to identify those areas that require more work, and determine, from Exercise B in chapter 2, which exercises in that chapter will help most		
BUILD CONFIDENCE	Confident in the presence of babies and children	Relaxed body language, investigative behaviour, and a willingness to approach	Relaxed body language, looking over at child; not actively seeking or wanting to approach	Tense body language, direct eye contact/staring. May bark in direction of the babies/children, and/or actively avoid
	Confident about sound/noise	Relaxed body language, investigative behaviour, and a willingness to approach	Relaxed body language, looking over at noise source; not actively seeking or wanting to approach	Tense body language, direct eye contact/staring. May bark in direction of the stimuli, and/or actively avoid
	Confident when touched and handled (all over, open mouth and gentle grasps by adult)	Will remain relaxed and calm, and not demonstrate any subtle signs of discomfort/calming signals. Will choose to stay	Will remain relaxed for a while, but if touched by certain people/in certain areas/certain types of touch (rougher grabs), or for a certain period of time, he may begin to show subtle signs of discomfort/calming signals and/or may start to try and move away	Will avoid being handled (may vary dependent on handler and could be total avoidance of touch/types of touch/locations on body
	Confident around moveable toys; baby walker, pram, etc	Willing to investigate the object; easily distracted by food/toy. Possible startle response; quick recovery time. Remains calm when object moving; willing to pass	Will take treats/toys in presence of object but looks over in direction of object. Takes longer to approach/investigate item; some signs of avoidance	Startled response; slow recovery time. Will not take food/toys. Nervous/reactive body language. Moves away from object whilst maintaining eye contact. Avoids object

Babies, kids and dogs

Walks/exercise	Confident around unfamiliar objects (high chair, basket, bouncer etc) and smells (nappies, lotions, talc, shampoo, creams, etc)	Willing to investigate the object, easily distracted by food/toy. Startled response, but is easily distracted by food/toy. Quick recovery time	Will take treats/toys in presence of object but looks over in direction of object. Takes longer to approach/investigate item	Startled response and slow recovery (if any). Won't take food/toys. Nervous/reactive body language, moves away from the object while maintaining eye contact. Avoids object
	On-lead walking (strength and behaviour), and behaviour when walking alongside a pram or bike	Is relaxed and walks calmly on a long, loose lead. Easy to control; no pulling. Will walk calmly next to a pram or bike, without pulling away from it	Does not pull constantly but is strong on lead. Does not walk calmly next to the pram and/or bike: pulls away or jumps about	Pulls a lot on the lead during walks; walks ahead for the majority of the walk; difficult to control and very strong on lead
	Walking past distractions (these can include dogs, other animals, cars, bikes and people. Do not put him in situations you know he will not be able to cope with in order to test this)	Walks past with ease without showing interest or at least stays calm generally, making it easy to pass by. Relaxed body language, investigative behaviour, but does not pull toward distraction	Walks along road, but shows more nervous/reactive behaviour when passing certain stimuli (eg cars). Easily distracted and often pulls toward stimuli	Barking, lunging towards stimuli throughout walk. Tense body language, direct eye contact/staring at stimuli
Flexible routine/energy	Energy levels (how much exercise is going to be needed)	Low level energy (fairly calm in most situations) – may need shorter walks through the day with extra stimulation at home	Medium energy levels (generally calm behaviour, but can have bouts of excitement/restlessness) – may need two-three walks per day with extra stimulation at home	High energy levels (does not often settle at home, seeks things to do/periods of excitement/ shows frustration at being bored). Requires minimum of two-three longer walks per day, as well as extra stimulation at home
Avoiding temper tantrums	Sits when greeted	Will sit to be greeted, and will keep all four paws on the floor. Generally calm and will just hover around nearby. Will approach calmy and will appear relaxed	Will sometimes jump up at people (possibly certain people or only when either out of house/in house). Can be quite excitable and is likely to knock over small children if they get in the way	Always jumps up at people to greet them, and will charge and lunge towards them. Very excitable and struggles to stay calm, possibly requiring holding back

Avoiding temper tantrums	Ability to give up or swap toys and play by himself	Will engage in play and easily give up the toy and settle at the end of the game. Will play with toys on his own around the house (not always seeking owner attention to play). Will readily release a higher value item for another	Will often initiate play with toys and can become overexcited. Will give up a high value item for another of same or higher value. Does not always understand when the game has finished and asks for more play.	Reluctant to allow anyone else to play with his toys. Can become frustrated if unable to get to his toys, or if they are taken away. Can become highly aroused during play sessions, which can sometimes lead to nipping. Reluctant to give up one toy for another
	Ability to leave food (high and low value)	Motivated by food (low and high rewards), but not highly excited by it. Will readily leave food, and will be calm around food	Motivated by higher value rewards. Excited by food/less inclined to leave it, but will still behave nicely and keep all four paws on the ground, and not try and snatch from hands	Extremely focused on food rewards. Reluctant to give up food/unable to leave it. When food is around, tends to lunge, jump, and try and grab out of hands
	Settles down when attention not provided	Happy to relax away from owner (on bed/ another room), and offers this behaviour of own accord	Will settle after a play/ walk away from owner. Can follow around house and seek attention occasionally through the day.	Follows owner around, constantly seeking attention. Reluctant to settle away from owner
To chew or not to chew	Taking and chewing inaproprate items (shoes, socks, food)	Doesn't take items when left out	Occasionaly takes items if left in a an area he is able to reach	Regularly takes items from all over the house; will run off with them and chew them until destroyed
Establish boundaries	Trainability and focus (train some basic behaviours)	Relaxed and calm in general. Listens to cues and is focused and keen to learn. Will work for food or toys; behaviours can be modified through training, therefore	Will focus, though can become distracted during training	Very difficult to get him to focus/change his behaviour, as either does not move or disengages. Requires a lot of high value reinforcement to maintain focus
	Behaviour when left alone for a short period of time	Relaxed and able to settle when left for up to four hours	Able to be left for shorter periods of time, though may start to become restless	Not able to be left for any length of time (whining, barking, destructive behaviour when this happens)

2: Preparation

The preparations you make in anticipation of your baby should involve your dog. His life is going to change a lot, after all, so it is crucial to help him adapt to having a baby or child around. Any new objects – such as baby equipment – should be introduced carefully and slowly, especially if he is nervous. The six areas noted below (and also in Exercise A: Assessment Chart, in Chapter 1) are those that require particular attention before your dog is around babies and children –

- Build confidence
- Walks/exercise
- Flexible routine/energy
- Avoiding temper tantrums
- To chew or not to chew?
- Establish boundaries

Preparing your dog

Exercise B provides guidance for dealing with the above six areas when introducing a new baby or child to your dog, and is also useful in situations where dogs are introduced into households where there are already children, and this is the first time that your dog will be introduced to them.

Completion of the Assessment Chart in Chapter 1 should have indicated how your dog coped with the stimuli/tasks related to Green, Amber or Red status for each, which will help you to understand his behaviour. Exercise B, opposite, details the training exercises linked with each stimuli/task, and indicates which exercises need more or less work, and in what order of priority. If, in Exercise A in Chapter 1, your dog scored amber or red for any tasks, the exercises for these areas should be prioritised, and more time allowed to work on them.

Use the exercises in conjunction with the notes about each of the six areas.

Build confidence

It is important for dogs to be confident and at ease with change, novelty, and most situations and sounds, in order to allow them to live safely and comfortably within our society, and be able to deal with new and strange situations and circumstances.

Confident dogs are generally those who have been well socialised to numerous objects, places, people, situations, and other animals, and can tolerate a variety of situations. If they are less fearful, they are generally more likely to be relaxed. Fear is a highly emotive behaviour, and, as discussed in Chapter 1, can manifest itself in many ways.

To be around children safely, your dog needs to be as robust and confident as possible, as children can bring a lot of uncertainty and unpredictability to his life. It's unrealistic to expect total harmony all of the time, however, and your dog should not *have* to be around children if he would rather not be. He should know that he can take himself out of a situation that involves children and babies without being pursued, and should be able to express his feelings calmly and be listened to. Consistency in this matter is key to ensure he remains relaxed, and does not feel the need for aggressive behaviour of any kind.

If your dog has not been around children before, or has lived for some time without children, there may be an increased risk of him becoming nervous or apprehensive in their presence. Another dog in the household can also affect how he may respond to new objects and situations, instilling him with confidence or anxiety, dependent on how the other dog behaves. It is therefore important to know how he

EXERCISE B

The six essentials	STIMULI/TASKS	PRIORITY	EXERCISE
BUILD CONFIDENCE	Confident in the presence of babies and children	■	1
	Confident about sound/noise	■	2
	Confident when touched and handled (all over, open mouth and gentle grabs by baby/child)	■	3
	Confident around moveable toys; baby walker, pram, etc	■	4
	Confident around unfamiliar objects (high chair, basket, bouncer etc) and smells (nappies, lotions, talc, shampoo, creams, etc)	■	4
WALKS/EXERCISE	On-lead walking (strength and behaviour), and behaviour when walking alongside a pram or bike	■	5
	Walking past distractions (these can include dogs, other animals, cars, bikes and people. Do not put him in situations you know he will not be able to cope with in order to test this)	■	refer to page 10
FLEXIBLE ROUTINE/ ENERGY	Energy levels (how much exercise is going to be needed)	■	refer to page 26
AVOIDING TEMPER TANTRUMS	Sits when greeted	■	6/7
	Ability to give up or swap toys and play by himself	■	8
	Ability to leave food (high and low value)	■	9
	Settles down when attention not provided	■	10
TO CHEW OR NOT TO CHEW	Taking and chewing inaproprate items (shoes, socks, food)	■	8
ESTABLISH BOUNDARIES	Trainability and focus (train some basic behaviours)	■	11
	Behaviour when left alone for a short period of time	■	12

KEY: ■ HIGH PRIORITY   ■ MEDIUM PRIORITY   ■ LOW PRIORITY

Babies, kids and dogs

will respond when with or without his canine companion.

In addition, if your dog is used to complete freedom within the house, a change to his daily routine – some restriction, say – will also require more preparation.

Once your baby arrives, or if your dog is around children more often, there will also be new and different objects, possibly more people around, and more (unfamiliar) noises. If sound sensitivity is a problem for your dog, you should address this issue as soon as possible, especially in regard to those noises that will be new to the household – a baby's cries, for example.

First exposure to *anything* new should be both controlled and positive. If you do not know whether your dog has had previous experience of any given situation, always assume that he hasn't, particularly with a puppy, as the younger he is the more variable his behaviour will be. An older dog may require longer to build confidence about things he may have developed a fear of. Conversely, if he is already quite confident and relaxed in novel situations and around new people and objects, less time may be required, although you should still introduce anything new in a calm, controlled and positive way to prevent initial nervousnous or anxiety.

All exercises should be carried out providing your dog with as much space as possible, so that he can move away if he wants to. As the object of the exercise is to build his confidence, if he does move away from an object that you are trying to acclimatise him to, start with more space between him and the object. The first few exposures should be calm and controlled so that nothing startles him and causes him to build fearful associations. Wherever possible, ensure there are no other animals present, as this could influence his behaviour.

The exercises within this chapter will help your dog to become more confident overall. Particular stimuli, for which your dog should be prepared, include the unpredictable movements of children, new sounds, touch, and new objects.

Unpredictability
Children can be quite unpredictable in their movements – erratic, fast, and invasive – with limbs all over the place, usually, which can be quite scary for dogs, especially if this is their first time around kids. If he does not understand that children can benefit and enhance his life – as providers of love, food and games – he can misunderstand their intentions, and his instinctive response would be to defend himself from possible harm.

In order to acclimatise your dog to the movement of children, complete Exercise 1 on page 30. Dogs can accept some of the hectic behaviours that children practice as long as their initial introduction to them has been positive.

Sounds
Babies and children introduce lots of new sounds, and some dogs can be sensitive to this. Exercise 2 on page 31 provides information on the desensitisation procedure; note you may need to begin at a lower volume level and increase slowly.

Contact: sudden touches and grabbing
Teach your dog to tolerate – and, hopefully, enjoy – positive touch, then teach your child appropriate direct touch. Continue to supervise interaction between them at all times.

Specific objects
Babies and children come with a lot of equipment (pram/pushchair, high chair, bikes, Moses basket, bouncer, doorway bouncer, baby walker) and toys, some of which move and are noisy. If he appears nervous around them and avoids getting too close, complete Exercise 4 on page 33. Those objects which move can cause the most fear responses, or even excitement. It is particularly important that he is confident approaching the pram/pushchair as he may need to walk alongside it.

Walks/exercise
His usual exercise regime may well change: in length, location; who is walking him and with what additional equipment. If he is already loose leash trained and will happily walk parallel to moving objects such as bikes, pushchairs and prams, not too much effort will be required, as these are the areas that are focused on in this section. Loose leash walking and the ability to walk alongside a pram/pushchair are the two main objectives.

Loose leash walking
The intention here is not to get your dog to stay by your side but to allow a walk that is enjoyable and relaxed for both

Stress-free walking on a loose leash makes for a safe outing that everyone enjoys ...

... though remember to always allow your dog the freedom of movement to investigate his surroundings: a sudden tug on the lead could mean you or the pram are pulled over.

Teaching your dog loose leash and parallel walking with objects is important when pushing your child's bike, or pram.

of you, allowing him the freedom to stop, sniff and wee, etc, but without pulling or lunging, which can pose a safety threat with a pram or children, or if pregnant.

Use a long lead (a training lead is ideal if you have one) of about six feet in length, but avoid using a flexi lead where possible. Always be aware of what he is doing and do not ignore him. Notice what he wants to do and respond appropriately, for instance, if he is nervous about another dog approaching.

When completing Exercise 5 on page 34, decide what is practical for you to do within the constraints of time and route availability, and determine what your dog's requirements are. If you are only able to meet his minimum walk requirements, be sure to provide mental and physical stimuli in other ways: two shorter walks rather than one long one, say, or games in the garden/food games, etc.

PARALLEL WALKING

Having built his confidence around the pram or pushchair, and taught him to walk on a loose leash, next comes training him to walk alongside moving objects on a loose leash. Start by getting him used to walking alongside an empty pram, initially. He should always be parallel to the pram or pushchair, with his lead loose, continuing loose leash training/reinforcement while introducing the pram. Once he is confident with this he should be able to transfer this skill to any moving object, such as a bike, as long as

Babies, kids and dogs

the new object is appropriately introduced. The main thing to be aware of is not to avoid running over his feet, which, understandably, could cause him to become nervous of the pushchair/pram/bike.

Flexible routine/energy

Once a baby comes along our priorities must change, which can upset your dog's usual routine: that 6am walk might have to be delayed if baby needs feeding, or he may not always get the long walk he is used to having every day. We have to teach him to be flexible, although it is, of course, essential to meet his needs, but if we do not prepare him for this, uncertainty about what's happening could lead to his becoming frustrated.

Providing your dog with something to do – such as a filled Kong™ or a chew – at various times of the day increases his flexibility, and keeps him entertained if you are otherwise engaged.

Some kind of routine for him (meal times, walks, etc) is essential so that he is able to predict when things are likely to occur, and can retain a degree of control over his day. Having a routine also helps you to meet his welfare needs, and you can throw in extra walks sometimes, provide treats, play games, or even do some training at different times.

Spontaneity and flexibility will not only provide him with mental stimulation, but will also allow for times of change when the usual daily routine is just not possible. Being used to change will allow him to adapt much better to situations such as this. Being able to cope with change and deal with inconsistency without becoming frustrated are important skills.

If it is necessary to change your (and his) routine in order to accommodate a new baby or child, ensure that this is done over a period of time. Try not to change all aspects at once (walk times, rules, sleep areas, meal times, etc) or too suddenly, and check that he is not just coping with the changes but is relaxed about what's happening (see Chapter 1 for relaxed behaviours).

Other changes that may or may not be required include –

- When feeding use a Kong™ or food toys and mazes instead of a bowl so that he is kept absorbed
- Adjust feeding/treat times
- Change the times/length/places that you walk
- Increase training (this may have already been implemented if following the exercises in this book) with new mental stimulation, new objects, and establishing boundaries
- Go upstairs alone at random times to help establish time apart
- Build in spontaneous play

Avoiding temper tantrums

Dogs can suffer from frustration, just like children, sometimes resulting in a temper tantrum. Although this does not sound too worrying, frustration – a highly emotional response – can become dangerous if it escalates, and should not be taken lightly.

Both children and dogs should be taught skills that allow them to deal with not always getting what they expect and want. Exercises 8 and 9 on pages 39 and 41 will help your dog control any frustration, teach him manners, and also how to settle himself and have 'quiet time.'

SIT TO GREET

Teaching your dog not to jump up at people he wishes to greet is very important, promoting his good manners and ensuring the safety of others. If he usually jumps up to greet *you*, you cannot expect him to know he should not do this to others, or when you are holding a baby or child in your arms.

First complete Exercise 6 on page 36 (unless he can already sit on request) before moving on to Exercise 7 on page 38.

SHARING

It is important to teach your dog to give up a toy without fuss, as this can prevent him becoming agitated when a resource is taken from him. By completing Exercise 8 on page 39, he will learn that it can be a good thing to share and give up toys.

NO SNATCHING

Waiting politely until given the go ahead to calmly and gently take food means that your dog is capable of self-control: an essential skill that can be useful in many situations. Teaching 'leave it' (Exercise 9 on page 41) becomes increasingly important as your child grows, moving around more and wanting to interact with your dog.

TIME OUT

Being able to entertain himself is another important skill for your dog to develop, especially when babies/children take priority, so teaching him to become more independent will be very beneficial for you both.

Initially, complete Exercise 10 on page 42 to teach him how to to settle, and then begin providing various types of mental stimulation to keep him occupied and entertained, such as treat games/puzzles, long-lasting chews, a stuffed Kong™, etc. Begin with easy-to-understand activities that will not cause frustration, gradually increasing the level of difficulty: for example, initially using small treats in toys, so that they fall out easily, but gradually increasing the size to make accessibility more difficult. It's vital that he understands how the activities work, as not doing so will also frustrate him.

To chew or not to chew?

Now that you have introduced the new items and objects that will become part of your dog's life, and he is confident around them, it's important to teach him not to chew them! If he does regularly chew furniture and household objects, this behaviour should be dealt with by a behaviourist, prior to the arrival of your baby. If, however, he does not chew persistently, once you have built his confidence (if this is required) and introduced the new items, you simply need to teach him that these are not to be chewed, and let him know which items he *can* chew.

This is achieved by using his toys during play, and

Toys left on the floor will be fair game to your dog – unless you teach him which things he can play with and chew.

having them available and accessible. Wherever possible, provide a variety of chewable items, and vary them, day-to-day. To teach him *not* to chew certain items, firstly distract him as soon as he picks up something that he should not. Tell him 'swap' or 'drop' (as with Exercise 8), replace the item with one of his own toys and have an exciting game with this to make it seem more interesting. The main objective of Exercise 8, the intention here is not to exchange the toys he should not be chewing with food, but with toys that he *can* chew, to teach him what he can and cannot play with.

He may need reminding, from time to time, which things he is allowed to chew, so be prepared for this. You can also use the 'leave it' cue, as taught in Exercise 9, if he is about to chew on something that he should not. Remember to reward him for leaving an item by picking up one of his toys and having a game with him.

Babies, kids and dogs

It's important to ensure that your dog receives not only exercise but also sufficient mental stimulation. Regular daily exercise should mean he will be more willing to settle, but also offer him mental stimulation throughout the day in the form of training exercises, or puzzle games/Kong™/puzzle feeder.

Establish boundaries

Boundaries allow your dog to understand how to behave appropriately within his home, and will prevent frustration at not knowing what to do. The rules of the house regarding furniture, garden, upstairs accessibility, etc should be agreed with all family members, and applied in a consistent way, helping him to understand what he is and is not to do. It's important that everyone sticks to these, as dogs are very good at discerning who might be a little lax in this respect ...

Certain steps can reduce risk around babies and toddlers. Baby gates can provide safe areas for both your dog and your child when you are unable to supervise interaction, allowing your child to move around safely and your dog to rest in peace.

Allowing your dog on the furniture is not a risk in itself, really, although can mean that he will be at face height with your child. And jumping off furniture or bounding downstairs can inadvertently result in knocks or falls.

As discussed in Flexible routines, resist making too many changes at once, and be mindful that boundaries are intended to provide a safer environment for all, and not as a way of isolating your dog. The promotion of positive interaction should still be the objective, with a means of safely separating child and dog during times of low supervision.

IMPLEMENTING HOUSE RULES

If something that your dog has always done becomes out of bounds once the baby or child is present, you will need to prepare him for the change in advance to allow him time to adjust.

Examples are if he will no longer be allowed on the furniture, sleep in your room, or travel in the front of the car. All family members must be in agreement in order to provide consistency. Exercise 11 on page 46 explains how to introduce an alternative behaviour.

CRATE OR BABY GATE?

Now is the time to gradually prepare your dog for changes to his boundaries, and areas he is allowed in/not allowed in, as well as acclimatising him to being alone for short periods. It is important that in all situations, he is sufficiently physically and mentally stimulated, so when you leave him confined within a certain area, say, or by himself, provide him with something to do, such as a stuffed Kong™ or toys.

Incorporate some of the exercises discussed in Avoiding temper tantrums: play with him within the area in which he is contained, play give and take, and use a Kong™ to practice a settle. When entering his space practice 'sit to greet.' Without stimulation of some kind, your dog could become bored and/or frustrated with being alone and confined. Establish boundaries but don't isolate him or forget about this member of your family. If you can teach him that being left is a positive experience, he will not see this as a punishment. Preparation is key. Invest in some Kongs™ and fill them in advance (see our suggested recipes in Chapter 5). Once he can easily access the food inside, try freezing them and serving in this way for extra challenge and stimulation (they'll also last longer). Have some Kongs™ in the freezer, ready for unforeseen circumstances.

To establish boundaries, you can use a crate (if crated as a puppy this might be a good option, as long as he is not left for extended periods), or introduce baby gates in some doorways to provide separation when needed. With either option ensure that the barrier is the right size for your dog. Puppies soon out-grow their crates, so a larger version will be required when he is fully grown; some athletic breeds can jump over standard-sized baby gates.

Should your dog become frustrated at being behind a barrier or separated from the rest of the family, he may demonstrate this by chewing or scratching the barrier, or the area around it. He also might vocalise a lot (bark and whine) particularly when he can see people on the other side of the barrier.

To try and avoid this occurring, introduce barriers slowly and gradually, and never use them to punish or isolate your dog. They are intended to help you supervise/control interaction only, and to use them in any other way is unkind. Exercise 12 on page 48 details the correct way to introduce a barrier.

Keeping your dog and child apart is important during training sessions, or when you cannot actively supervise interaction between them.

Take great care not to isolate your dog from the rest of the family, though, as this will negatively impact on his behaviour. At best, he will feel lonely and depressed; at worst, he may feel the need to resort to attention-seeking behaviours, such as barking, whining, scratching, etc.

29

This exercise is intended to acclimatise your dog to the erratic and sometimes unpredictable movements of children, and should be carried out slowly and carefully. Take care not to overwhelm your dog with lots of small children at once, and do not place a child in any situation where your dog could react adversely.

Where possible, carry out the exercise in an area that is generally calm and not too busy, with sufficient space to allow a distance between your dog and the children to prevent any feeling of confinement; therefore reducing the chances of him having a bad experience. Throughout the exercise give your dog a lot of praise to provide positive reinforcement for being in the presence of children, and to establish an association between them and rewards/nice things.

- To begin with start exercising him more in areas where children can be found, the local play park, for example, walking some distance from the play area to begin with. Observe him for any signs of stress. If he looks comfortable and relaxed, gradually decrease the distance to the play area until he is happy to walk/sit there while children are playing.

- If children should walk past him while you are out, ask him to sit or stand still, rewarding him with treats when he does. Allow him to look in the direction of the children, rewarding calm and relaxed behaviour.

- Once confident that your dog is relaxed around children and the noise they make, progress to allowing him to have more contact with them. If you have friends or family with young children, ask if they would be willing to meet your dog and you. This should be done in surroundings that are familiar to him and where he feels comfortable, to reduce the risk of his feeling anxious or threatened. It may be that his own home is where he will feel most at ease, as long as he is able to distance himself if he feels the need to.

- Start with older children who are able to follow instruction (they should be given guidance and supervision at every stage). If the exercise is carried out in your home, ask the children to enter the room quietly and calmly, and sit on the sofa. Ideally, they should carry out an activity – drawing or reading, say – whilst you observe how your dog reacts. Should he appear uncomfortable or stressed at any point, ensure he is

This dog confidently approaches the baby and toy, showing very calm, relaxed body postures and movements.

able to remove himself from the situation, should he choose to do so. On no account try and force him to interact: building his confidence in situations such as this may take time and lots of positive encouragement. If he chooses to approach the children for a sniff, allow him to do so, and softly praise and reward him for any positive response.

- If he is comfortable with the situation and happy to stay close to the children, allow them to stroke him gently and quietly, avoiding any sensitive areas (as discussed in Chapter 5).

- Once you are completely happy that your dog is comfortable with this situation, you can begin to introduce him to younger children in the same way, ensuring that he is not in a position where a bad experience could result.

- If your dog *should* have a bad experience with a child (because of a sudden noise or movement, say), defuse and manage the situation by either having a game with him to take his mind off the experience, or going back a few stages in the exercise.

- If unable to provide children to carry out this structured greeting, while you are socialising your dog around children's play areas, etc, and if he is comfortable and relaxed, allow older children to approach and stroke him. Have treats ready to reward him with, gradually reducing these as he becomes used to the experience. Do not allow any child to approach him if he is at all nervous or agitated: go back to a point where he was most comfortable and start again from there, building confidence slowly.

Exercise 2: Desensitise to sound

Your dog may have already received some exposure to various sounds when out on walks or during Exercise 1. But if you are having a baby, or he is particularly sensitive to sound, using specialist CDs or apps of recordings to introduce various noises and sounds in a more controlled manner. Those containing new household noises and sounds (baby cries, whimpers, laughing, etc), could prove particularly helpful.

- Begin with the recording at a very low volume, played in the background during the day (but not continuously). Continue intermittently at this level over a few days.

- If your dog shows relaxed behaviours, and does not even seem to acknowledge the noise, increase the volume slightly. He may look in the direction of the noise but should still be relaxed and not be too bothered about it. If he does become anxious and uncomfortable, reduce the volume until he is again comfortable with it, then introduce an increase more gradually.

- Increase the volume just a little each time, ensuring that your dog remains relaxed, until it is at a level commensurate with what can be expected from a baby.

It's vital you introduce your dog to a baby's cries and the usual sounds of children gradually and carefully, so that these do not alarm or frighten him.

Exercise 3: Handling protocol _____

Teaching him to tolerate – and enjoy – being touched can be incorporated into fun games that you'll both enjoy.

- Get down on the ground with your dog, and encourage him to come to you. Reward him when he does so with a few treats and praise. When he has settled into a comfortable position, gently touch him in non-sensitive areas of his body, while rewarding him with more treats. Your touch should be firm enough for him to feel. Initially, provide the treats with one hand while touching him with the other, but then try touching him first and rewarding him after, so that the treats are not a distraction but a reward for him remaining calm whilst being touched.

- If he remains calm and relaxed, begin to gently touch him in more sensitive areas of his body, such as his ears, paws and tail, as these are the parts that children are most likely to grab. When doing this you may need to use more treats at first, and gradually reduce the number given as he gets used to the contact. Continue to praise and reassure him verbally.

- If he is comfortable with this level of contact, very gradually increase the firmness of touch. Do not use too much pressure. Remember to intermittently reinforce this exercise.

It is important that your dog is happy to be touched and handled all over, especially in sensitive areas that a child may grab, including his ears.

Gently touch your dog's paws while rewarding him for calm, relaxed responses.

The main aim of this exercise is for your dog to feel confident around new items and objects, without paying them much attention, and this is achieved by rewarding him for approaching and calmly sniffing the objects to reassure himself that there is nothing to worry about. Do not use too many rewards during the exercise, especially if he is already fairly confident. Treats should be given to reward an approach to the objects, and not placed on them, as this could cause him to perceive the object as being a source of food. Reward him when he approaches the object, but do not force an approach or continuously reward him when he is around the object.

   If he is nervous, introduce things slowly or at lower levels of activity and/or volume.

Allow your dog to approach and investigate the new object in his own time.

- Place the object in the corner of a room or outside with lots of space around, where your dog can approach and sniff it without feeling confined. As he approaches the object, drop treats near to it and in front of him to encourage him to continue. Reward calm approaches such as sniffing.

- When he is confident about approaching the object and is generally relaxed around it (refer to Chapter 1 for relaxed body postures), move to the next step.

- Place the object in various locations over a few days so that he gets used to the changes, and is still happy to approach it.

- Place the object where he will need to pass it in closer proximity (the aim is that he should simply walk past the object without paying it any attention at all).

- If the object in question moves or is noisy (a musical/movable toy or Hoover, for example), introduce this somewhere that he can move away from it. Have a helper move the object a little (but not directly towards him), with someone else dropping treats close to the object (but not too close if he is not confident about approaching the object).

Gently drop the treats around the object to encourage him to approach it. He should then associate it with good things.

continued overleaf ...

Always closely observe your dog's body language, checking for calm and relaxed postures and expressions.

- Build his confidence around the object before introducing/turning on the sound, doing so very gradually so that he remains relaxed, regardless of what the object is doing or the noise it is making.

- If he is startled by any object or noise, drop some treats on the floor around him so that the object or noise does not acquire a negative association. Resume the introduction at a later date at a further distance or lower volume.

- Be on the lookout for signs that your dog is trying to avoid the object, and introduce a distance from it by providing more space or slowing the process.

- Be aware that a moving object can sometimes motivate dogs to try and prevent them from moving, or they can become aroused/excited, and want to chase or play with the items. Move them slowly initially and gradually increase the movement, or else only allow your child to play with such items out of sight of your dog.

Exercise 5: Teaching a loose leash walk

Top tip

- Try to avoid giving him cues. The objective here is for your dog to understand what he needs to do, so chooses to do it himself, rather than being asked

- Always keep the leash loose

- Decide on which side you would like your dog to walk. Find a quiet spot in which to work.

- Tempt your dog to sit or stand alongside your leg, and feed him treats so that he associates being in this position with receiving a reward. This will require practice until he is happy to remain by your side on a loose leach for prolonged periods, without the need for treats.

- Ensure that the leash is slack, with no pressure on your dog's collar or harness.

Begin by rewarding your dog for being next to your leg. (You may need a lot of rewards to establish this positive association.)

- Take a step forward, and, as your dog moves with you, feed him a treat.

- If he keeps pace with you as you slowly walk forward, feed him a couple more treats.

- After practicing this for a day or two, ask that he walks a little further before receiving a treat.

- If he walks ahead of you, or pulls in any other direction, immediately stand still, say nothing, and wait for his attention to come back to you. You will need to be patient at this point and avoid tugging on the lead, or getting his attention by calling him. Ideally, you want him to choose to come back to you.

- Once you have his attention again, show him a treat in your fingers to return him to your side, and walk a couple of steps before giving the treat to him.

Move forward just one step, rewarding your dog for moving with you.

As you take more steps, you can begin to reward less often, but make sure that you gradually reduce the treats, and keep it varied.

continued overleaf ...

- Do not feed him a treat as soon as he comes back to your side as this can soon teach him that the quickest way to get a treat is to pull away and then return.

- Repeat the above in different places, and gradually introduce various distractions. Practice this until he is able to walk by your side for 15-20 paces without pulling before giving him a treat.

- By this stage in the training, most dogs will find that moving forward is a more important reward than food, so you can begin to fade out the treats, though still use these in places where there are many distractions to help him get it right.

Top tip

- Do not wrap the lead around your wrist whilst walking, because if your dog should make a sudden lunge, you could be injured, or you and the pram could be pulled over.

- If you are practicing a loose leash walk alongside a pushchair/pram/bike, ensure there is a comfortable distance between this and your dog so that he does not feel threatened by it.

Make sure the leash is always loose. If he should pull, stop and wait for him to come back before again moving forward.

Exercise 6: Teaching a sit

- To begin with ask your dog to focus on you by showing him you have a nice treat in your hand.

- Hold the treat in front of his nose and slowly raise your hand up over his head.

- His eyes (and head) should naturally follow the direction of your hand, and his bottom should automatically begin to lower.

- When his bottom touches the floor reward him with the treat.

- Practice this a few times, until his bottom always touches the floor.

- When he is doing this reliably each time, begin to add in your verbal 'sit' cue just before you move your hand up over his head.

Hold the treat in front of his nose. and move it up and back over his head ... as his head and gaze follows the treat in your hand, his bottom should automatically lower.

- Once he is focusing on your verbal cue, use the physical cue less often (your hand going up), and simply ask for a 'sit' (but do not try this too soon).

- Once this has been achieved, extend the time between his sitting and providing the reward to increase the length of the sit (this may help when teaching Exercise 7).

TOP TIP
- Remember to keep training this until it becomes automatic for him.

- Remember to provide top-up training intermittently throughout his life so that the behaviour remains reinforced.

When his bottom touches the floor, give him the treat in this position.

37

Exercise 7: Teaching an appropriate greeting ————

Greeting people on walks

Ideally, you will have the help of two or three people to practice this exercise. If you are pregnant and training this, we recommend that someone else does the greeting first, so that your dog will learn and understand what he should do when greeting people, and especially not to jump up at a pregnant woman.

- Have your dog on a lead to begin with in order to retain control over the situation. Just before you meet the nominated 'visitor', ask him to sit, using your verbal cue 'sit' as previously trained in Exercise 6.

- Keep treating him while he remains in the sit position. Allow the 'visitor' to approach and offer your dog a treat, or provide him with some attention/say hello, dependent on what your dog is comfortable with.

- Repeat the above sequence until your dog will automatically offer a sit when greeting people without requiring the verbal cue (the action has become habitual and the presence of a person becomes the new cue).

- Gradually reduce the number of treats you give him as greeting the person becomes the new reinforcement. However, you should still reinforce the behaviour with treats now and then.

Coming through the door

As in the above exercise, your dog should be able to calmly greet you and your baby/child when you return home. Have the help of at least two people to practice this exercise, and ensure the person coming through the door knows not to touch or talk to your dog initially unless he is already sitting.

Ask your dog to sit before you open the door. If he remains sitting as the visitor comes through the door, reward him with a treat and ask your 'visitor' to greet your dog. If your dog jumps up at the visitor, he should ignore this behaviour

Ask your dog to sit calmly. Show him a treat in your hand to encourage him to remain sitting.

Open the door and reward your dog with a treat for maintaining the sit position. Shut the door if he rises and begins to move forward.

Ask the 'visitor' to greet him calmly while he is still in the sit position.

and slightly turn away from your dog. Regain your dog's attention and ask for a 'sit', then reward. It's important that your dog does not learn the chain behaviour of jump-sit-reward, so you may need to work on the 'sit' request a little longer, asking him to wait before getting his reward.

Also practice this while opening the door (with no-one there) without the greeting, so that he gets used to sitting when he sees someone or when the door opens.

Exercise 8: Sharing toys

- Prior to beginning this exercise have a variety of high and low reward treats available. It may be necessary to offer a high reward treat to encourage your dog to let go of something, especially if he is initially reluctant to give up his toy.

- Have a toy you know he likes to play with, and that you know he is able to get a good grip of.

- Get on the floor with your dog, and encourage him to have a good game with you. Once he is engaged in the game and has hold of the toy, offer him a treat (holding it close to his nose so that he can smell it).

- Once he lets go of the toy give him the treat. After he has eaten the treat begin the game again. (If he is

Hold one end of the toy in one hand while presenting a treat in the other ...

continued overleaf ...

... and reward him when he lets go of the toy. You can also use this manoeuvre to swop something he shouldn't have with something he can have.

reluctant to release the toy, simply hold the toy still so that the game stops. Offer a higher value reward to encourage your dog to release the toy.

- Once he is consistently releasing the toy on presentation of a treat, add in your verbal command (eg 'swap' or 'drop') JUST BEFORE you present the treat, so that he listens to the command and is not distracted by the treat. Continue to practice this with him on a regular basis.

- When you want to end the game, stop playing with the toy and give your dog a verbal command to signal to him that the game is over. This command could be the same as that given in Exercise 10: Settle ('off you go').

TOP TIP

- If at any point during play he inadvertently nips or grabs your hand, stop the game and put away the toy. Your dog should learn that mouthing means that play will stop. If he randomly mouths at other times, refer to Chapter 1 for guidance on how to deal with this.

Exercise 9: Teaching 'leave it'

- Have a tasty treat in your closed right hand and hold this near to your dog's nose to allow him to smell it.

- When he stops sniffing your closed right hand, bring this arm down to your side, and reward him with a treat from your left hand. If, after moving your right hand to your side away from him, he is still interested in it, still give him the treat from your left hand, keeping your right hand closed.

- Repeat these steps until he is reliably moving away, then add your 'leave it' cue, repeating several times over a few days.

- Next, you can add in your 'take' cue just before you reward him from your left hand (this is not essential so only train it if you feel it will be helpful to yourself and him).

- Now begin to increase the time he stays away from your right (closed) hand by offering more treats from your left hand while he is calm and waiting patiently, and is not trying to get the food out of your closed right hand.

- Then, begin to open your right hand while he calmly chooses to stay away (reward him with a treat from your left hand). If he approaches your right hand, close it and wait for him to sit and patiently wait before trying again.

- Once he is not approaching or showing any interest in your right hand when it is open, swop to a higher value reward (increasing the level of difficulty and temptation), while still giving treats from your left hand, using the 'take' command.

Hold the treat in your closed hand near his nose so that he knows it's there.

Once your dog has an understanding of the 'leave it' command, you can use this in other situations, too. Try placing a treat on the floor or on a low table, and asking him to sit at a distance. As you position the treat, give the 'leave it' cue. If he chooses to leave/stay away from the treat, reward this behaviour by giving him a treat from your hand.

Once he is doing all of the above consistently, you could increase the difficulty by asking him to walk past a pile of tasty treats, or pop a treat on a table without asking him to sit first. Remember, though, that the more successful he is, the more likely he will be to repeat the behaviour: set him up to be successful and don't ask him to do anything too difficult.

continued overleaf ...

Once he stops showing interest in the treat in your right hand ...

... reward him with a treat from your other hand. Practice this until you are able to hold open your right hand with the treat in it, whilst feeding him a treat with yuor left hand.

Canine 'sharks'

If you know that your dog can be quite impulsive, and may accidently nip your fingers when you are rewarding him, to avoid this, always offer the treat on an open palm, or hold the treat with your fingertips, with only a tiny part of the treat poking out, so that he learns to be more careful and precise about taking the treat (this also allows you to keep hold of the treat until he is calmer and being more careful).

During this exercise he should learn to be calm, and should begin to curb his impulsive behaviour. Try and avoid giving him the treat if he lunges for it, otherwise this uncontrolled behaviour is being rewarded.

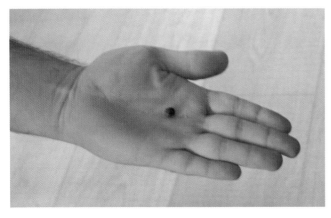

Offer treats on a flat, open palm with fingers held together ...

... or hold the treat with your fingertips, to avoid your fingers being nipped inadvertently.

Exercise 10: Teaching a settle

- Before beginning this exercise ensure that your dog has had a walk or a game to expend any excess enery.

- Place his bed in a quiet area (ideally, where it will be located once the baby arrives).

- Encourage your dog over to his bed with treats. Scatter a few on the bed, drawing him onto the bed until he has all four paws there (he does not need to be sitting or laying down at this point).

- Continue feeding him the treats on his bed, and he should decide to get more comfortable by sitting or laying down. Disengage from him as much as possible by not talking to him, touching him, or giving eye contact, although still carefully observing his behaviour to ensure you are rewarding him sufficiently for him to want to remain there.

- While he is calm and relaxed, keep dropping treats at random on his bed. You will have to drop treats regularly to begin with to encourage him to stay on the bed, but as he progresses, increase the time between dropping the treats (therefore encouraging a longer calm and settled period).

Encourage your dog onto his bed by scattering treats there ...

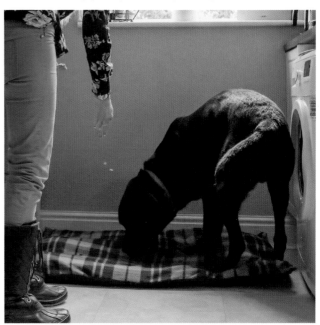

43

continued overleaf ...

... and continue to scatter treats randomly until he decides to sit or lay down on his bed.

While he is calm and relaxed, keep dropping treats at random on his bed.   *Substitute treats for a stuffed Kong™ or tasty chew to keep him content on his bed.*

- Have a variety of treats available, and always try to reward more for when he sits or lays down of his own volition, without the need for a cue of any kind.

- When you would like to end the training (keep sessions short to begin with, ending before you anticipate he might move away). Give him a release request – this can be verbal: for example, 'off you go,' or physical (scattering a few treats away from the bed). Once you have given this release cue, do not give any more treats.

- After a few minutes, he may return to his bed in the hope of getting more treats, in which case reward the behaviour by dropping a few treats nearby to encourage him to remain in place for longer and longer periods.

- Once the behaviour becomes established, and he is settling for long periods, reduce the number of treats and add in verbal rewards ('good boy!') to reinforce the behaviour once the treats have been phased out.

Top tip

- You can substitute treats for a stuffed Kong™ or tasty chew to keep him occupied on his bed.

- Be consistent in your cues – decide as a family what these will be and ensure you all use the same ones.

Exercise 11: Introducing an alternative behaviour for a new rule

Remember that your dog will not understand straight away that he is no longer allowed to do certain things; things that he has always done in the past. To succeed in implementing new rules and teaching an alternative behaviour, it is important that you and the rest of the family remain patient and consistent with him during this time of adjustment.

- Begin by first training and then reinforcing the new behaviour (settle on his bed; not going upstairs).

- When he goes somewhere he no longer should, or exhibits behaviour you no longer want, using a treat or a toy, encourage him into an area he is allowed to go, closing off access to where he was (see Exercise 12), if possible, so that he is not tempted to return, or ask him for a new behaviour he has learned, and then reward him/play with him for complying. Don't do this too often, however, as he may just learn to do these things as a way of getting a reward.

- If he continues to exhibit unwanted behaviour, you may have to physically prevent his being able to do this by blocking access to the sofa, say, or using a baby gate to prevent access.

- Always praise and reward him for making the correct choices and choosing to perform the new behaviour.

If you no longer want him to go on the chairs, first get his attention by showing him a treat ...

... and use this to guide him off the furniture (you may need to use a high value reward initially).

Once he is fully off the chair, praise him and give him the treat. Remember that, previous to this, you should have taught him how to settle on his bed, so that he knows he has somewehere to go that is comfortable and safe.

Whether using a crate or baby gates, the procedure for introducing them is similar.

- Ensure that the area/place your dog will be confined to is sufficiently spacious, comfortable and draught-free. Provide a comfortable bed, water, and some toys. Leave the gate/door of the crate open, and encourage him to enter and investigate whenever he wants, rewarding him when he does so.

- When your dog is relaxed and happy about entering the crate/area by himself, you can begin to close the gate/door, initially for just a moment, after which scatter some treats on the floor and re-open the gate/door. Repeat a few times, gradually increasing the time that the gate/door remains closed.

- If he remains relaxed within the area, the next time that you close the gate/door, leave him with something that will occupy him, and move away from the area. Very gradually leave him for slightly longer periods, returning to open the gate/door and leave it open, and at other times just get on with household chores, etc, leaving the gate/door closed.

- If your dog shows signs of distress or frustration at any point, go back several steps and train this exercise more slowly.

Provide your dog with a space that is comfortable. Give him good things whilst he isin there, so that he chooses to go into of his own volition.

When he is calm and relaxed, and absorbed with his toy or Kong™, begin to close the door for short periods of time – but immediately open it again should he show signs of distress or frustration.

Top tip

- The best time to train this exercise is when your dog has been on a long walk, and is ready to relax and rest.

- Do not leave him for too long, to the point where it becomes a negative experience, or he has to get your attention to let you know that he needs to toilet.

- Remember to spend time with him in this space, and do not isolate him, or shut him out of family life for long periods.

3: Introducing the new family member

Arriving home with a new baby is an exciting time for all involved. It is natural to be concerned about how your dog may react to the baby, but if you have put in the work from Chapters 1 and 2 of this book, you will have helped prepare him for this moment. The following guidance details how to introduce your baby to your dog for the first time in a safe and positive way.

Protocol for the first few meetings

Step one: Greet your dog
Greet him on your own as he is likely to be excited to see you, and may jump up, especially if he has not seen you for some time. Have treats ready to reinforce appropriate greetings. If he is overly-excited to see you, move out into the garden with him for a play session or some other form of attention before he meets your new baby.

Step two: In the house
Ask your dog to settle before you bring the baby into the house – refer back to Avoiding temper tantrums in Chapter 2.

Someone (this could be you, or, alternatively, someone who is known to him, and with whom he feels comfortable but is less likely to jump up at) should bring the baby into the house, whilst another person he knows calmly strokes him and talks to him, ready with a treat or toy to distract him if necessary.

Step three: Approaching and greeting the baby
To ensure that this is a positive experience for your dog do not force him to interact with the baby, but, with you holding the baby, allow him to sniff the baby's feet, or the blanket or cot to investigate the strange, new smells. Praise him for calm, quiet behaviour, and also when he moves away from the baby.

If you are concerned that he may jump up, attach his lead and allow it to trail on the floor behind him, so that you can stand on the lead instead of grabbing at his collar if you feel he is becoming over-excited. If this does not result in him becoming calmer and more controlled, guide him away from the situation, using treats, if necessary.

The most important thing is for you, your dog and the situation to remain calm and relaxed, with behaviour as normal as possible. If it is safe to do so, allow him to sniff the baby, and, even if you do not normally like dogs to lick people, understand and take into account that this is normal greeting behaviour for dogs that will help him to feel in control. Should he do this and you don't want him to, on no account should you reprimand him, or startle him by exclaiming, say. Simply call him away and distract him with another activity or a treat. Throughout the entire process of your dog greeting your baby talk to him calmly, and have whoever is holding the baby gently stroke him with their free hand. Make every greeting a positive experience for your dog, and whenever he is around your child, and, eventually, you'll probably find that he will be completely at ease with the new presence.

Step Four: Attention away from your baby
To end the initial interaction between your dog and your baby, calmly call your dog away from the baby and reward him with some quiet attention, to avoid arousing him too soon after the interaction.

Use the steps above to gradually acclimatise your dog to the baby coming and going within the house, until he becomes so used to this that he may well just ignore it. Always ensure that there is someone around to supervise interaction between the two: never leave them on their own together.

NERVOUSNESS

This issue should, ideally, have been dealt with prior to your baby's arrival, but if your dog is still nervous it is really important to ensure that he can always move away from any situation that he finds uncomfortable when going through the above introduction. Always proceed slowly, and on no account try and force interaction between your dog and your baby (or anyone else, for that matter). If you are at all concerned about your dog's behaviour around your baby or child, consult a qualified clinical behaviourist for guidance.

Going forward
DIRECTED CONTACT

Directed contact is where a child is able to touch/stroke your dog with guidance from you. Demonstrating a flat hand with which to touch your dog to preclude your child from grabbing his coat or body (his nose, say), guide your child's open hand, gently keeping their fingers spread to prevent them grabbing his hair (babies naturally close their hands into fists and clutch anything that touches them). Should this happen, keeping the baby's arm outstretched so that he cannot actually pull on the hair, as gently and quickly as posisble, open your baby's hand to release the hair. Praise and reward your dog for remaining calm and relaxed.

WEANING

Dogs usually quickly associate children with food as they very often carry food around, and/or drop this from highchairs or onto the floor. Your dog quickly makes a positive association with the child as a result, which is great for building his confidence. The danger with this is he may gain access to food that is not good for him (and even dangerous, as in the case of chocolate, say), and overfeeding will result in unwanted weight-gain. As your child grows, he could also reinforce begging behaviour in your dog, as well as other unwanted behaviours (jumping up at a high chair, for example), so it's best to separate the two when your child is eating.

Problems may also occur once your toddler is able to

This kind of close attention can be extremely worrying for a dog and should be avoided at all costs.

move about by himself, when he may deliberately give your dog food. Always supervise when there's food around, and endeavour to teach your toddler which foods are bad for

Babies, kids and dogs

Show your child how to hold his hand open and flat to stroke your dog, preventing him from grabbing at his hair.

This is a much more appropriate interaction, with the dog looking comfortable and at ease. Children should be encouraged to stroke dogs in this way.

Although this is positive interaction, there is a danger that the dog may mistakenly nip the baby's fingers when trying to get at the food she is holding.

Large dogs can thieve food from highchairs, so take care to supervise, or separate them during mealtimes.

your dog. Hopefully, having completed Exercise 9, your dog should be able to leave food when requested, and gently and politely take appropriate food that is offered to him.

Floor play

When you and your children are playing on the floor, try and ensure your dog also receives attention and, where safe to do so, include him in what you are doing. Have him settle on his bed with a Kong™ in the same area, for example, or provide toys for him to chew, or simply just talk to him during your play so that he feels part of it. Make sure that his physical and mental needs have been met beforehand, though – he has been walked, had a game of ball in the garden, done some training, etc, to reduce the chances of him becoming over-excited whilst your toddler is on the floor.

53

Babies, kids and dogs

If your child plays on the floor, ensure that your dog has his own space, and is not crowded out or compromised in any way.

Ensure that during floor playtime your dog has enough space to move away from your baby or child if he wants to.

Your dog can learn a lot about babies and children by simply observing them at play.

Remember: never punish or reprimand your dog for showing interest in your child. Should he start to become excitable or over-aroused, however, you can redirect his behaviour onto something else or distract and guide him away calmly with treats. Avoid him becoming over-excited by keeping arousal levels low, and actively supervise at all times.

Children of any age are not known for being quiet, and your dog should have received earlier preparation for the different sounds he may be expected to now hear. However, even in an ordinarily calm dog, change can cause uncertainty, so be aware of this and always continue to observe and gauge his behaviour for the subtle signs of discomfort. Children become scarier and more unpredictable to dogs and animals in general as they begin moving around, and, as they grow, the noises they make will change and sound different.

Also keep in mind that your dog will be aging along with your child, and pain associated with old age (arthritis, etc) becomes more likely. Even though he may be fine with being handled, sudden, acute pain may cause him to react unpredictably, so ensure he has regular checks-ups with his vet. Regularly handle him and look him over yourself to check for signs of ill health, as well as reinforce that he should remain calm when touched in possibly sensitive areas. If you notice any changes in your dog's behaviour, such as unsure looks, avoidance, movement problems, etc, seek your vet's advice. If it turns out not to be a medical concern it may be that more training is required: perhaps revisit Exercise 3 Handling protocol.

Important points
- Your dog should not be expected to put up with being grabbed or mishandled, lots of noise, or having things taken from him.

- Physical punishment is cruel, unkind and unnecessary, and will cause your dog to become fearful and anxious; possibly resulting in a dangerous situation. **Never use it.**

- Do not always assume that your dog is at fault if an incident develops. It's quite possible that your child may have caused him discomfort or pain, or done something that has made him feel threatened or fearful.

- Respond immediately to subtle signs of discomfort from your dog to prevent a situation escalating, and to let your dog know that he can trust you to act accordingly, and in his interests.

- Should your dog have to resort to growling for whatever reason, calmly and carefully remove your child to a place of safety, and then try and determine what it is that has made your dog feel that growling was his only option in the circumstances.

- Should your dog pick up one of your child's toys, ensure that it is you who retrieves it from him, and not your child (refer to page 27).

4: Toddler training

Now that your dog has learned to develop the skills he needs to behave appropriately around children, it's time to teach your children how to behave appropriately and interact safely with dogs.

This chapter is designed to help you help your child gain an understanding about dog behaviour, and how to behave around them, using 'Champ and Charlie' illustrations.

Champ   Charlie

Toddlers, by their very nature, can be difficult to manage around dogs, and doing so safely requires increased vigilance and patience. Toddlers move differently to others, and their behaviour can be unpredictable: they're also noisier and louder, generally – all of which can be very scary for some dogs.

Statistics show that children under five have an increased risk of being bitten by dogs, often by animals who are known to the child, such as the family dog or a grandparent's dog. And they are often bitten around the face, which could be due to the dog being a similar height to the child, and the fact that children tend to get very close to whatever they are interested in. Young children cannot read a dog's body language as easily as someone older, and a number of studies have demonstrated that children are unable to correctly interpret a dog's facial expressions: they can perceive a growl or show of teeth as a dog 'smiling,' for example. It is therefore essential to teach children how to understand canine behaviour and behave appropriately around dogs.

The 'Champ and Charlie' illustrations demonstrate child/dog interactions, and should be used to instigate discussion, appropriate to your child's age and level of understanding, to teach him how to behave around dogs, and understand why this should be the case. Each illustration suggests questions and points to discuss with your child, to prompt him to think about how and why it is important to interact with a dog only in certain ways.

Happy dog vs unhappy dog
Some excellent resources are available to parents about how to educate children to be around dogs, and these include –

- The Blue Dog (http://www.thebluedog.org/en/)
- The Kennel Club Safe and Sound (http://www.thekennelclub.org.uk/training/safe-and-sound/)
- Doggone Safe (http://www.doggonesafe.com/)

These programmes help children understand how to correctly interpret canine behaviour, and how to make the right choices around dogs in various everyday situations. Using these resources will enable you to help your child recognise when a dog is relaxed and comfortable with a situation, and when he is saying that he has had enough.

Toddler touches
Toddlers will often grab and pull at a dog as a way of getting his attention, showing him affection, or in an attempt to play, and will need to be taught the correct and safe way to touch him (refer to Directed contact in chapter 3 for more information), which will also include when to touch him

56

continued on page 60

How do we know Champ is unhappy?

Turning his head away

Looking from the corners of his eyes

Corners of his mouth would be tense

Movements would be slow

Avoiding being touched

Champ is displaying body postures and movements that let Charlie know he has had enough interaction, or does not feel comfortable. If Champ also growls or shows his teeth, it's important that Charlie understands that this means he has definitely had enough, and it is time to give him some space to do what he wants. Champ should never be punished for growling, as this is simply his way of communicating that he is not comfortable.

How do we know Champ is unhappy?

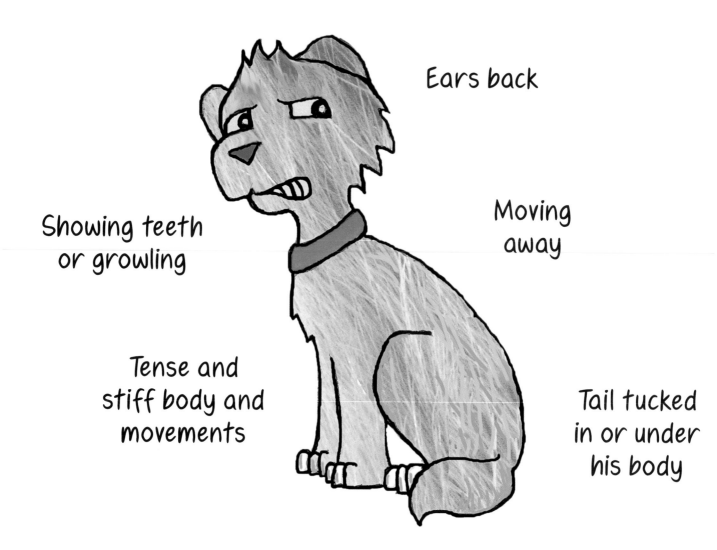

Ears back

Showing teeth or growling

Moving away

Tense and stiff body and movements

Tail tucked in or under his body

How do we know Champ is happy?

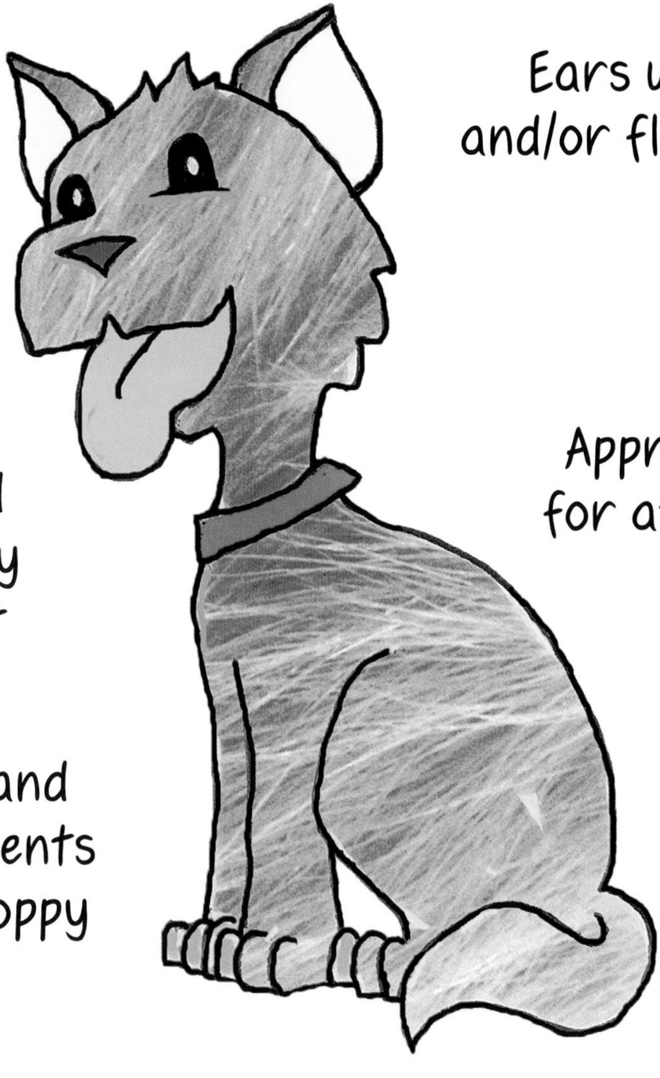

Ears up and/or floppy

Not growling or showing his teeth

Mouth relaxed and tongue may be hanging out

Approaches for attention

Legs and movements are floppy

Tail wagging and relaxed

When Champ is relaxed and enjoying an interaction, his body movements will be loose, and he will approach Charlie with floppy ears and tail.

Babies, kids and dogs

and where. A good way to involve your toddler with correct handling is to have him help when grooming your dog, although you must have done Exercise 3 Handling protocol, in chapter 2, first. This assumes, of course, that being brushed is something your dog enjoys.

Toddlers will often grab at faces, paws, ears and tails, and will also climb, sit on and pretend to ride dogs. This can be dangerous to a child, because, when attempting to climb on a dog they could be thrown off if the dog should move away, and the dog may respond negatively to your child attempting this. In addition, allowing a child to mistreat a dog in this way is wrong: he is not a toy and should be treated with kindness and respect at all times.

Respecting your dog's food/resources

As discussed previously, you are responsible for supervising any interaction between your child and your dog, and never leave your child and dog alone together. This is especially important when your child is likely to be inthe vicinity of your dog's food and/or other resources. To minimise risk, arrange to give your dog his meals when your toddler is not present (napping), or at least not free to move around (in his high chair).

Toddlers, naturally, will be interested in everything that your dog does, and, if not supervised, may try and take food from your dog's mouth, play with his toys, play with his food bowls, put their hands in his water, and lie on his bed (his place of safety and security), all of which can, quite understandably, cause your dog to become frustrated or defensive, or to feel threatened by the loss of what he considers to be *his* resources, with unwanted results. Always ensure that his bed and resources are in an area of the house away from young children: he should be able to completely relax without fear of being disturbed.

PREVENT YOUR CHILD FROM –
- Hugging your dog
- Grabbing your dog's hair
- Touching and grabbing your dog's paws
- Pulling your dog's tail
- Pulling your dog's ears
- Climbing on top of/sitting on your dog
- Pursuing your dog to try and interact

ENCOURAGE YOUR CHILD TO –
- Gently stroke your dog on his chest, back or near his shoulders
- Brush your dog under supervision, and in the sure knowledge that he likes this
- Interact when your dog chooses to approach

PREVENT YOUR CHILD FROM –
- Taking your dog's/his own toys from him
- Taking food from your dog/touching his food bowl
- Sitting/playing in your dog's bed
- Teasing your dog with food
- Feeding your dog something that could make him ill

ENCOURAGE YOUR CHILD TO –
- Eat his food away from your dog
- Leave your dog alone when he has high value resources such as a bone
- Play appropriately with your dog with his toys under supervision (ie a game of 'fetch')
- Safely offer appropriate treats to your dog when he is calm, under supervision

Let sleeping dogs lie ...

A dog who is sleeping should not be disturbed, as this can startle him, which he may perceive as a threat. If your dog chooses to go and settle (on his bed, for example), he should be allowed to rest in peace, and your child should accept that he should be left alone. This could be an ideal opportunity to enjoy some floor play with your child, or for you and your toddler to fill a Kong™ for when your dog wakes up.

PREVENT YOUR CHILD FROM –
- Being around your dog when he is sleeping
- Waking your dog when he is sleeping

ENCOURAGE YOUR CHILD TO –
- Engage in alternative activities when your dog is sleeping: reading, floor play, etc
- Prepare some new resources or food toys for your dog for when he wakes

Does Champ look happy or unhappy – why?

Why do you think Champ could be worried?

Why should Charlie not pull Champ's fur?

If Champ walked away or growled, what would you do?

Why should Charlie not play with Champ's paws?

How can you tell that
Champ is not happy?

How could Charlie touch Champ to make him happy?

Could this hurt Champ?

What is Charlie doing to Champ?

How can you tell that Champ is happy?

How is Charlie touching Champ?

Why is this a nice way to touch Champ?

Why do you think Champ is unhappy?

Would you be upset if somebody teased you like Charlie is teasing Champ?

Why would Champ be unhappy when Charlie is on his bed playing with his toys?

How can we
tell Champ is
unhappy?

What could Charlie
do to share?

Why should Charlie never take food or bones from Champ's mouth?

When is it okay to hold and play with Champ's food and toys?

What do you think Charlie is doing for Champ?

How do you think Champ feels?

What are Charlie and Champ doing together?

How would you play
with Champ?

Why do you think they are
both enjoying this game?

How is Charlie waking up Champ?

Do you think
Champ is
unhappy?
– Why?

What is Charlie doing?

5: Raising children around dogs

Unfamiliar dogs

Not surprisingly, more care and supervision is required when children are around unfamiliar dogs, whether inside or outside the home, as the dogs may not have been through the same socialisation and training process as your own dog. Chapter 4 provided information about appreciating and understanding when a dog should not be disturbed, and also how a dog should be touched when interacting with him. There should also be a recognition and understanding of when a dog looks uncomfortable, or is giving subtle warning signs to stay away. These skills should now be combined with an understanding of when and how to greet a dog for the first time.

A dog can become nervous around children and people he doesn't know, and in such a situation his owner, the parent and child should carefully observe his behaviour at all times to check for any discomfort. This is also very relevant when introducing a dog into the family home, which can be quite a stressful time, particularly as resident dogs and animals can intensify the nervousness of the newcomer. It's strongly advised that an assessment of behaviour, as described in Chapter 1, be completed.

Familiar dogs

Do not allow your child to invade a dog's space, and ensure that they always behave in a calm and consistent way around them. You cannot control the environment beyond your home as much as you can within it, and so, if your child does go somewhere where he will meet other dogs, make sure that you do the following prior to this happening –

- Ensure you have met any unfamiliar dog, and been around him during interaction with children – and specifically your own child – in order to assess his behaviour in these situations.

- Check how the dog's owner/carer intends to supervise his dog around your child when you are not present.

- Remind your child of the appropriate behaviour around dogs, and especially mention that they should never enter the property or garden of where the dog lives if he is there but the adults are not around.

If you are not happy about the answers you receive to these three points, talk with the dog's owner/carer and ask whether alternative measures can be put in place. If this is not possible, decide whether or not you want your child placed in such a situation ...

Also keep in mind that when around any dogs – even your own – in different environments, new stimuli can cause them to feel different emotions, such as anxiety and frustration, especially if they have not been through the confidence-building exercises.

Top Tip

- If your dog begins to behave aggressively around children, for whatever reason, tell the children to stand perfectly still and quiet with their arms folded, and not to look at the dog to prevent the situation escalating. Quickly and calmly take your dog out of the situation to a place where he feels safe and comfortable. Try to determine exactly what it was that caused him to behave in this way, and take steps to prevent a recurrence.

- A growl is good: this is a warning that a dog is feeling

Babies, kids and dogs

OBSERVE
Ensure your child follows these steps when greeting an unfamiliar dog –

Owner's permission	Ask the dog's owner/carer whether the dog would be happy to be touched and stroked
Be approached	Only engage with the dog if he willingly approaches first, as this is an indication that he wants to interact
Sniff hand	If he is relaxed and calm in his approach, offer an open hand for him to sniff
Engage interaction	If he remains calm and relaxed after this, he can be gently stroked on the side of his neck, on his side, or on his back. Do not place hands over/on his head (which can be construed as threatening), or hug him anywhere
Relaxed	Continuously observe the dog to ensure he remains in a relaxed state. If at any point he tenses or looks unsure, stop touching him and allow him to move away
Voluntary interaction	Any interaction should always be voluntary on the dog's part, and he should actively choose to accept the attention
End of engagement	Should/when he moves away, do not pursue the dog or force attention on him. Break contact, and only provide him with more attention if he approaches again

uncomfortable, and is not simply aggressive behaviour. Never ignore a warning of this sort. Animals will choose to avoid aggression, given the chance, as it comes with attendant risk of injury, and uses energy unnecessarily.

● Never leave your children on their own with a dog/dogs. Always be aware of what children and dogs are doing, and if you or another responsible adult cannot be present, ensure that they are safely and securely separated.

As children grow, they are likely to interact more frequently with your dog, so continue to observe his behaviour, particularly as he will also be growing older. Prevention is better than cure, and even the sweetest-tempered dog can have a bad day, or experience pain, especially as he ages.

Bringing a new dog into the home
If you already have children and are thinking of bringing a new dog into your home – and regardless of whether or not you already have a canine family member – there are a few things that you should do first –

PREPARE THE HOME
Make your home a comfortable place for your dog to be, with an area arranged just for him. Be ready with prepared

Always ask for permission to approach/touch a dog first ... ... and only stroke him if he approaches with relaxed body postures.

When the dog moves away, cease the interaction and do not pursue him.

rules and boundaries so that these can be consistently applied from day one. Allow him time to settle in.

The right dog
Give careful thought to which is the right type of dog for your family, your household environment, routine, and experience. This does not mean choosing a specific breed/crossbreed, although be aware of the traits and requirements of particular breeds. And, of course, larger/ stronger breeds/crossbreeds have the potential to the cause more damage due to their size and bite strength. All dogs, regardless of size, breed or type, can pose a risk – even one that is life-threatening – so never be complacent with those breeds/crossbreeds that reputation indicates might be calmer or 'safer'. Remember, too, that bites from small dogs, whilst common, are less often reported in the press.

Educate
If this is the first time that a dog has been part of your family,

Babies, kids and dogs

Help your pup to settle into his new home by providing him with comfortable, draught-free areas to relax in, and sufficient mental stimulation.

ensure that everyone has the necessary knowledge about what's involved in caring for a dog, and is fully prepared to do so. As already mentioned, it's especially important that your children know how to behave when interacting with dogs, and, likewise, your dog should be taught appropriate behaviour using positive reinforcement (rewarding correct behaviour). Even if you have owned dogs before, every one is an individual, and some require more guidance and reinforcement during training than others.

Settle in

Let him settle in in his own time, and be patient. Encourage positive interaction using treats and toys, and build his confidence with every exposure to new situations and objects. Let him have some control over his environment by allowing him to approach and investigate in his own time. If he is particularly young or particularly nervous, he may require more formal confidence-building exercises and training (see Chapters 2 and 3). Don't overburden him with too much at once, and try not to make a fuss if things do not always go to plan. Allow him to adjust to what is a new routine for him in his own time, and if you have a multi-

animal household, gradually introduce them to each other in neutral and controlled areas.

Get your children involved

Children can benefit greatly from growing up with dogs and other animals. Dogs can provide important health and social benefits, as well as loyal friendship, and can teach children much about compassion and consideration for other species. As your children grow up alongside their canine companions, they will learn respect and responsibility, as well as empathy for animals and people. From toddler age they can begin to interact and become involved with him in some safe activities, and as they get older, they can be given more responsibility when caring for him, taking into account their ability and knowledge, and your dog's temperament.

Basic training

Children can often get involved in many aspects of your dog's daily training routine, which can help establish a bond and positive relationship between them. A particularly nice training exercise to begin with is for your child to help teach him a recall when in the garden. This is a fun exercise for your child to do, and should be done in addition to formal recall training (which should be taught with a different cue word). A good option would be for your child to use your dog's name when training this exercise rather than the cue you use to recall him.

The following exercises are intended as guidance for the trainer (your child), although there are also important points for you as parent and supervisor.

SIT AND WAIT

Teach a sit as outlined in Chapter 2 (Exercise 6) if he does not already know how to do this.

- Place your dog in a sit (hold the treat above his head and slowly take it backward until his bottom touches the floor, then have your child give the treat).

- Take one step away from him. If he stays where he is, return and treat.

- Repeat this several times, and once you feel confident that he will stay, add in your 'wait' cue. Gradually increase

1) Show your child how to hold a treat in his hand, and slowly raise it up and over the head of your dog.

2) As your dog's head and eyes follow the treat, his bottom should automatically lower toward the ground.

3) When your dog is in the sit position, your child should reward him with the treat.

the distance you move away from him, but set him up to succeed by not going very far from him to begin with. He needs to understand what you are asking him to do before increasing the difficulty.

- This exercise, along with the recall (below), gives a nice sequence/trick for your child to do.

RECALL

- Have a selection of tasty treats ready before you start. Let your dog wander around the garden, or ask him to sit and wait if he can do this.

Babies, kids and dogs

Tell your child which cue words he can use, which should sound as exciting as possible to your dog.

Either ask your dog to sit and wait, or simply let him wander in the garden.

- When he is a short distance away, have your child call his name.

- When he responds by coming to your child, encourage him to sit and have your child reward him.

Note: Show your child how to hold the treat in a closed hand and take it close to your dog's mouth, opening out their hand flat to allow him to take the treat. If you know that your dog sometimes mouths when taking treats, have your child drop the treat on the floor (while you work on this behaviour separately: see Exercise 9 in chapter 2). Repeat the above steps until he is correctly responding to your child on a regular basis.

Exercise
Children generally like to be involved in walking their dog, especially if they are able to hold his lead. In order to do this

Your child should call your dog to him – making it sound as exciting and inviting as possible – and provide him with the treat when he responds. If he can ask your dog to sit before doing this, even better!

safely, train your dog in loose leash walking prior to allowing your child to take his lead, so that you can be confident he will not pull or behave erratically. Refer to chapter 2 'Walks' for guidance or a recap on the training.

Attach two leads to his harness: one for you and one for your child to hold. In this way, you retain control. Keep an eye on the tension of the lead, and ensure it remains loose at all times. Discourage your child from wrapping the lead around their hand, as there is more chance they will

be pulled over, should the lead be tugged or jerked for any reason. If you do involve your children in this way, it may be a good idea, initially, to let your dog run off any excess energy before you attach another lead for your child to hold. This will make it easier for both you and your child, and give your dog an opportunity to toilet.

Games
One of the best ways to get your child interacting with your

dog in a positive way is to teach appropriate games to play with him.

Although tug games can be used effectively in training, and are good to play with adults, if your dog should become too excited he may inadvertently begin to mouth. It may be best to avoid playing games of this sort with children, for this reason, and instead try fetching games, or hide-and-seek games.

FETCH THE BALL

Make sure that he has learnt to give the ball safely (teaching a give-and-take). If he has previously shown signs of resource-guarding, this will need to be worked on before allowing your child to play with him with the ball.

Also be aware that some dogs can unintentionally knock over a child, so it is a good idea to teach him to sit and wait patiently (page 76). Teach your child not to tease your dog with the ball, or wait too long to throw it for him, as this can cause frustration on his part.

HIDE-AND-SEEK (TOYS AND TREATS)

Hide-and-seek is a fun game that children of all ages can play with your dog. The idea is that the children hide treats or toys around a room or in the garden while your dog is elsewhere, and then he is allowed to search for and find the treats. Show your dog, first, how the game works, by letting him see the treats and that they are being hidden in the chosen area, otherwise he may have no idea of what he is supposed to do. Your child can encourage your dog into different areas using hand gestures to provide guidance to him.

As before, this is not a suitable game to play if your dog has shown signs of guarding food or toys.

Trick training

Trick training is a fantastic way to get your child involved with your dog, as this is fun, and is often preferred to obedience-style exercises.

Following are a few simple tricks that you can help your child teach your dog, and show them off to family and friends.

Dual lead walks mean that your child can get involved more actively, while you can still maintain safe control.

Playing 'fetch' is a safe and enjoyable activity for both your dog and child. Always check your dog's arousal levels, however, and supervise young children.

TRAINING A SPIN *(SEE PAGE 82)*

- Get your dog's attention by holding some nice treats in front of his nose, then slowly move your hand to the right (or left – whichever is your preference).

- As he moves his head to follow, keep moving your hand until you have completed a full circle around his head. Your dog should turn and follow as you do this. If he

is reluctant to begin with, break down each part to make it easier for him to do as you ask. As he moves his head to follow, reward him with a small treat, then move your hand further round and do the same again. Continue in this way until he completes a full circle (spin).

- Once he has mastered several full spins with a reward, begin to add in your cue (most dogs will respond better

Show your dog a treat in your hand, and slowly move your hand to the right or left – whichever is your preference.

As he moves his head to follow, keep moving your hand until you have completed a full circle around his head.

Your dog should turn and follow as you do this.

As he moves his head to follow, reward him with a small treat, then move your hand further round and do the same again. Continue in this way until he completes a full circle (spin).

Once he has mastered several full spins with a reward, begin to add in your cue word.

Playing tricks provides skills and mental stimulation for both trainer and dog, and, more importantly, establishes bonds.

to a hand signal than a verbal cue, so use your hand to signal a spin, but not all the way around his head, just in front of him.

- You may find it easier at this stage to show your dog the treat in the hand that you're not using to signal a spin. Then reward him from this hand once he has completed the full circle.

GIVE A PAW

- Ask your dog to sit, and let him know you have some tasty treats in a closed fist.

- Hold this hand close to one of his paws.

- He is likely to use this paw to touch your hand, as he knows the treat is there, and of he does, give him the treat. If he doesn't do this, try touching his paw with your other hand to see if he will lift it.

- Continue to do this until he raises the same paw reliably. You may need to encourage him to raise his paw higher each time by waiting for a slightly higher paw lift each time until you reward him.

Note: If you wish – and are confident he will be happy to do so – you can allow your child to hold out his hand and 'catch' your dog's paw when he lifts it up.

When he is reliably lifting his paw (8/10 times) you can then add in a 'give paw' cue.

TARGET CONE TOUCH

Target training can be a great trick for any dog (but particularly those who know a few tricks already), and is an excellent way to teach a dog to focus. It can be very useful when a dog's becoming overexcited, or when there's the possibility of a reaction to another dog in the environment. A small cone is ideal but most object you might find in the garden will suffice: a turned over plant pot, for example.

- Get your dog's attention, and place the cone (or other object) close to him. If he investigates this, mark the behaviour by saying 'good boy' as soon as he is close to or touching the object, then reward him with a treat.

Encourage your dog to come to the target by pointing at it, or showing him a treat in your hand ...

... and, as he leans forward to touch the target, praise him and give him the treat.

Rip up lots of cardboard and other packaging, scrunch it up and place it all within a larger box ...

... add various types of dog treats to the box ...

... mixing them up in the box and making sure they're well-hidden ...

... then show your dog the box - and watch him enjoy hunting through the paper to find the treats!

If he seems to need a little encouragement, try tapping the object.

- Repeat this several times.

- Once he is reliably touching the object, add in your cue, such as 'touch cone', saying this just as he's about to touch the object.

- To make it more challenging, gradually move the object further away from him, until he has to cross the garden to touch it and return for a treat.

Messy play
BOX OF TREATS

Your child will love this, but your dog should definitely love it more! Save up some cardboard boxes or shoeboxes and newspapers or toilet tubes in order to make your box of treats.

- Tear up old newspapers.

- Fill the box with the shredded paper or toilet tubes.

- Hide some of his favourite treats or normal kibble inside the box amongst the paper and tubes.

- Give him the box and watch as he has loads of fun pulling out all of the paper, etc, to find and eat his treats.

Note: Then all you have to do is clear up the mess (sorry!).

ICE CUBE TRAYS

A simple, perfect treat for your dog on a nice day. Your child will love watching him throw and chase these icy treats around the garden.

- Place a treat in each cube in the tray, top up with water and freeze – easy as pie.

- For a tasty (and healthy) change, try using chopped apple or carrot instead, or fill the cube with fruit juice or yogurt. Especially good on a hot day.

Provide your child with all of the ingredients required ...

... and let him place a dog treat into each ice cube space within the tray.

Babies, kids and dogs

Next, pour water into each ice cub within their tray to surround the treats ...

... and pop in the freezer for several hours. A crunchy and novel reward for your dog, especially on a hot day.

STUFFING KONGS™

This is an excellent way in which to involve your child, which will help him learn about the importance of stimulation for your dog. Stuff a Kong™ with your chosen ingredients (avoiding anything that is unhealthy or potentially toxic), then either give the Kong™ to your dog straight away to settle with, or freeze it for an extra level of difficulty, or to have available at busy times.

A few recipes for you to try are included at the end of this chapter, or you can simply use your dog's usual food (soaked in a little warm water if it is dry). Or try topping off the Kong™ with a little soft cheese or paté.

Gather the ingredients together and begin spooning them into the Kong™ ...

... filling it right to the top, pressing down as you go.

The Kong™ is then ready to give to your dog ... or you could pop it in the freezer for a couple of hours so that the filling will last longer.

86

Babies, kids and dogs

Note: Be absolutely sure that the ingredients you use are safe for your dog (and your child). It's advisable to reduce your dog's usual food allowance if feeding Kongs™ in addition to this, otherwise there may be a tendency for him to gain weight, particularly with high-calorie fillings such as peanut butter. If you feed a Kong™ *instead* of a meal, this does not need to happen.

Kong™ recipe ideas

FILLING RECIPE FOR MY DOG'S KONG™

BASIC

- Soak some of your dog's usual kibble in warm water and allow to cool

- Let your child mash the kibble with a fork

- Mix in some tasty treats of your choice or, for more fun, have a variety available, from which your child can create unique combinations (or you can simply follow the suggested recipes)

- Fill the Kong™ with a teaspoon or fork, pushing the filling right to the bottom

- Top with soft cheese or pate (or inroduce a layer of this halfway up the Kong™) to make the treat even more enticing

- If you're not intending to give your dog the Kong™ straight away, make sure he doesn't see you filling it or he will be very disappointed not to receive it then

FILLING RECIPE FOR MY DOG'S KONG™

PEANUT BUTTER BISCUITS

Ingredients
Your dog's usual kibble
Smooth peanut butter
Honey
Fish oil (optional)

Preparation method
- Place his kibble in a plastic bag and allow your child to bash to bits with a rolling pin

- Once crushed, pour the kibble into a bowl and add a teaspoon of honey

- Mix in a tablespoon of peanut butter and a dash of warm water, ensuring that all of the kibble is coated

- Add a dash of fish oil if wanted

FILLING RECIPE FOR MY DOG'S KONG™

KNICKERBLOCKER GLORY

Ingredients
Low-fat natural yogurt
Strawberries
Apple sauce (ready-made from a shop is okay, though you could make your own with less sugar in it)
Banana

Preparation method
The best thing about this recipe is that you can use most fruits (but NOT grapes, raisins or sultanas, and be aware that the seeds of peaches, pears, plums, apricots and apples all contain arsenic ...) You can even allow your child to tuck in!

*   Have your child mash or chop (dependent on their age) the different fruits

*   Begin stuffing the Kong™ with a spoonful of mashed banana followed by a spoonful of yogurt, then some mashed strawberry, more yogurt, then apple sauce, topped off with more yogurt

*   Repeat as necessary until the Kong™ is filled

*   For best result freeze for an hour or two before giving to your dog

FILLING RECIPE FOR MY DOG'S KONG™

TURKEY SUPREME

Ingredients
3oz of turkey mince
¼ cup of brown rice
Pinch of dry rosemary
One cup of water

Preparation method
*   Rinse rice with cold water

*   Brown the turkey mince, then add the rice, together with a dash of dry herbs and the water

*   Bring to the boil and simmer until the rice is tender

*   Drain excess water and allow to cool before stuffing the Kong™

FILLING RECIPE FOR MY DOG'S KONG™

SUNDAY BEST

Ingredients
Leftovers from your Sunday dinner, including –
Mashed potato
Vegetables (NOT onions or garlic)
Meat

Preparation method
Use up any unwanted Sunday roast leftovers in this quick and easy recipe.

- Chop the vegetables and meat into bite-size pieces and mix well with the mashed potato

- Stuff the Kong™ and let your dog enjoy straight away, or freeze for a tasty treat later

Note: Gravy is high in salt and not recommended for dogs.

FILLING RECIPE FOR MY DOG'S KONG™

MAC AND CHEESE

Ingredients
Pasta
Grated Cheddar cheese (or soft cheese)
Various vegetables

Preparation method
- Cook the pasta as per the instructions and allow to cool

- Mix in one tablespoon of cheese

- Add the cooked or raw vegetables and mix well

Appendix: resources & further reading

Resources
Doggone Safe and the 'be a tree' programme
 http://doggonesafe.com/

The Kennel Club Safe and Sound scheme
http://www.thekennelclub.org.uk/training/safe-and-sound/

The Blue Dog
http://www.thebluedog.org/en/

The Blue Cross
https://www.bluecross.org.uk/pet-advice/be-safe-dogs

Dog Bite Prevention in Children
http://www.vet.utk.edu/dogbiteprevention/parents/index.html

Family Paws Parent Education
http://familypaws.com

Further reading
The Canine Commandments • Kendal Shepherd • Broadcast Books
Uses humour and compassion to teach positive interactions with dogs

Life skills for puppies – laying the foundation for a loving, lasting relationship • Helen Zulch and Daniel Mills • Hubble & Hattie
The key skills a dog needs, and how to build a fulfilling relationship

Visit Hubble and Hattie on the web: www.hubbleandhattie.com
www.hubbleandhattie.blogspot.co.uk
• Details of all books • Special offers • Newsletter • New book news

Index

Aggressive behaviour 9, 12, 22, 23, 62, 73, 74
Arousal 9, 14, 15, 17
Attention-seeking 8, 9

Boundaries 21-23, 28, 29, 48, 75

Canine Ladder of Aggression 12
Chewing 21-23, 27, 28
Cognitive decline 11
Confidence 19, 22, 23, 31-33
Conflict anxiety 9
Cues 8, 21, 27, 28, 34, 37, 38, 40, 47, 76

Emotions 7, 8
Energy 20

Fear 9, 11, 13
Flexible routine 20
'Focus on me' 10-11
Frustration 9, 14, 15, 26-28, 48, 60, 80

Games 79-81
Getting a new dog 74-76

Kong™ recipes 88-90

Meeting a new dog 73-75
Messy play 84-90
Mouthing 14, 15, 40

Nervousness 22, 24, 51, 73

Observational skills 15-21

Pain 9, 11, 18, 55, 74
Play 13-15, 26, 28, 29, 40, 53, 54, 70
Predatory 8

Reactivity 10, 11
Recall 77-79
Resource guarding 13, 14, 80
Routine 22, 23

Temper tantrums 20, 22, 23, 26, 28
Temperament 7, 17
Touch/handling 11, 19, 32, 51, 56, 60, 62, 63
Trainability 21
Trick training 80-83

Self-control 27
Settle 21, 76
Sit 21, 37, 38, 76
Sound sensitivity/noise 19, 23, 31, 33, 54
Stimuli 17, 21-24
Stress 9, 12, 13, 16, 19-21

Unpredictability 24

Walks/exercise 10, 11, 20, 22, 23, 25-27, 35, 36, 78, 79
Warning signals 9, 12, 13, 16, 19-21

More great books from Hubble & Hattie!

HOME ALONE – AND HAPPY!
Essential life skills for preventing separation anxiety in dogs and puppies

Kate Mallatratt

Hubble & Hattie

HELPING MINDS MEET
Skills for a better life with your dog

Helen Zulch & Daniel Mills

Hubble & Hattie

 This unique book, written by professionals in the field, explains how and why misunderstandings occur between us and our canine companions, and how we can work to resolve them. It aims to help us adjust the way we interact with our dogs, in order to help our dogs be well behaved, whilst at the same time enabling us to enjoy fulfilling relationships and a good quality of life with our canine companions

205x205mm · 96 pages · 100 colour images · paperback plus flaps · ISBN 9781845845766 · £12.99*

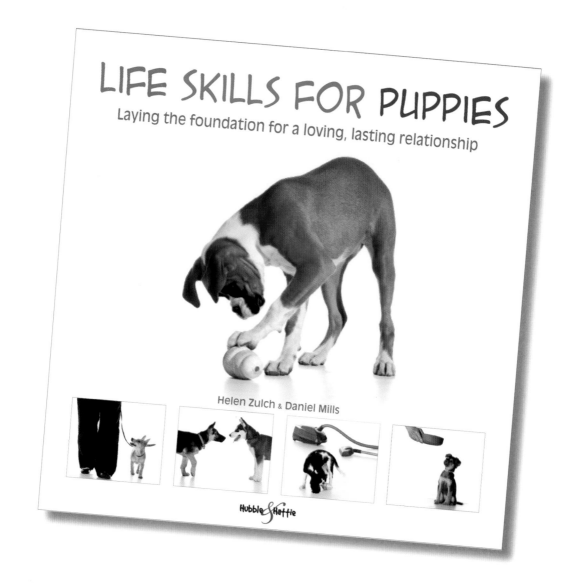

LIFE SKILLS FOR PUPPIES
Laying the foundation for a loving, lasting relationship

Helen Zulch & Daniel Mills

Hubble & Hattie

Puppy education from the puppy's perspective! Presenting the key skills that a dog needs to cope with life, this ground-breaking book, written by professionals in the field, aims to assist owners develop a fulfilling relationship with their puppy, helping him to behave in an appropriate manner and develop resilience, whilst maintaining good welfare. The skills taught are incorporated into everyday life so that training time is reduced, and practising good manners and appropriate behaviour become a way of life.

205x205mm • 96 pages • 121 colour images • paperback plus flaps • ISBN 9781845844462 • £12.99*

Your dog and you ...

Understanding the canine psyche

Gill Garratt

Hubble & Hattie

This ground breaking book examines the relationship between people and dogs from a psychological perspective, uniquely incorporating Cognitive Behavioural Therapy (CBT) to facilitate this. A dog's behaviour can be a reflection of the emotions an owner may be experiencing; it follows that insight into our behaviour using CBT to reduce emotional unrest will, in turn, be reflected in our dog's behaviour. Dogs have had to become experts at reading people in order to live with us. By understanding our dog and how he responds to us, we can comprehend more about our world and how our dog sees us. Dogs are naturally expert psychologists, and have, over centuries, been bred and domesticated to live harmoniously with us. That they have – in the main – achieved this so well reflects this amazing animal's ingenuity.

205x205mm · 96 pages · 99 colour images · paperback plus flaps · ISBN 9781845847388 · £12.99*